General CHEMISTRY Laboratory II

CHEM 117 LAB MANUAL

Concepts and Experiments

Second Edition

Yan-Yeung Luk
Syracuse University

Cover photos courtesy of the author.

Kendall Hunt
publishing company
www.kendallhunt.com
Send all inquiries to:
4050 Westmark Drive
Dubuque, IA 52004-1840

ISBN 978-1-4652-1210-8

Printed in the United States of America
10 9 8 7 6 5 4 3 2 1

Contents

Dear Students

The second term of general chemistry laboratory (CHE 117) focuses on understanding different concepts in general chemistry through hands on experience in performing experiments. As such, this lab manual emphasizes on understanding rather than elaborate procedure for doing experiments. To achieve this purpose, this lab manual contains two new and original experiments: one resolves understanding symmetry and chirality and the other introduces contemporary understanding of salt dissolution in water. The manual also contains eight redesigned LABS that are revised and modified from traditionally used experiments for this course.

Besides introducing the idea and practice of doing experiments, one major goal of this course is to prepare students for advanced courses that many of you will take in the coming semesters. The LABs of symmetry and soaps help prepare students for organic chemistry, the LABs concerning liquid crystals and dissolution of salts are some basic topics for materials sciences and physical chemistry, the LABs involving topics of equilibrium and kinetics are helpful for analytical chemistry that are related to forensic sciences. This lab manual is written aiming to be self-sufficient, but the schedule of the experiments will still accommodate the progress in the lecture (CHE 116). That means while the dates for having laboratory will not change, the exact experiments may be switched for different dates. For any of such switching, you will be notified at least one week ahead of time.

Please always check the following website for details:

http://blackboard.syr.edu

The website and emails will communicate any last-minute changes. And you can always contact your teaching assistants for questions.

The lab manual for CHE 117 contains a PRE-LAB section, which lists questions pertaining to the experiments you will perform for a particular laboratory, and a POST-LAB section, which contains a list of questions pertaining to the experiments you have performed. Both the PRE-LAB and POST-LAB sections can be torn off, and you must complete the PRE-LAB questions and hand it in before doing the experiment for that day, and hand in your answer for the POST-LAB questions at the beginning of the lab for the following week. For both PRE-LAB and POST-LAB, if needed, please attach additional pages for answering the questions.

This manual is still a work in progress that may contain errors and unexpected problems although the experiments have been test-run several times. The teaching assistants and I will do our best to resolve problems as quickly as possible. Please let the teaching assistant or myself know if there is anything unclear, so we can improve on this manual.

Thank you.

Yan-Yeung Luk

Chemistry Department

Notes for the Teaching Assistants

Grading for the Last question in the postlabs.

	CASE A	CASE B
Students write down "Nothing"	10	0
No answer	10	0

Case A: Student scores more than 90 points for the rest of the postlab.
Case B: Student scores more than 90 points for the rest of the postlab.

SAFETY PRACTICES

Syracuse University

Accidents in a chemistry laboratory can happen and can cause serious and sometimes permanent injuries or even death. But **all accidents can be avoided**! In this class, you must be alert at all times, maintain a safety-conscious mindset, follow the rules and regulations, and exercise common sense.

A chemical laboratory contains many potential hazards. It is your responsibility to learn about them, about how to exercise precautions, and know the proper actions if an accident does occur. In general, you must

1. Read the experiment before coming to the laboratory.
2. Use common sense when working with chemicals and apparatus. Know how to get help should an accident happens.

Once your work area has been assigned, find and note the nearest location of the safety items listed below. Keep the list with you during each lab session. Quick, common-sense action can avoid a serious accident. Teaching assistants (TAs) are assigned for each section consisting of 18–20 undergraduate students. Please pay close attention to your teaching assistant (TA) when he/she points out the locations and explains the use of the following items. Together with other itmes, there is quiz at the end of this check-in section.

Emergency phone	Fire alarms
Emergency exits	Fire extinguishers
First-aid station	Fire blankets
Eyewash stations	MSDS sheets
Emergency showers	

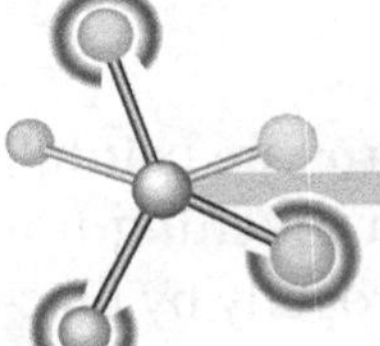

Before Starting the Experiment

1. Come properly attired. Shorts, short skirts, bare feet, sandals or open shoes are not allowed in the chemistry laboratory. Tie back/pin up long hair. Do not wear long necklaces or dangling jewelry, loose or bulky clothing, or expensive clothing. Clothing can be, and has been, ruined by the chemicals in the labs. NOTE: You will be dismissed from class and marked absent if this dress code is not followed.

2. Pay close attention to the instructions from your teaching assistant (TA). The TA will provide valuable information for the experiments, including safety measures, and may change some aspects of the experiment. Listen to his/her pre-lab announcements and ask questions if you do not understand the instructions.
3. Read and understand all the specific safety precautions of each experiment with careful attention to all hazard warnings. Ask your teaching assistant if you have questions about any aspect of the experiment. You will be given information on the location of the Material Safety Data Sheets (MSDS) for chemicals in use in the lab. MSDS information includes physical properties; health hazards; reactivity, fire, and explosion data; procedures in case of spillage; disposal information; and any special precautions. Ask your TA if you have any questions regarding the MSDS.
4. Write a note to inform your instructor and TA of any known allergies or medical conditions that might require you to limit exposure to chemicals. Discuss with your TA any questions regarding allergies and medical conditions that might raise concerns about working in the lab.
5. Keep personal items off the bench top; besides creating a hazard, they can be contaminated by chemicals.
6. Eating, drinking, and using cell phones are not allowed in the lab at any time. Turn off cell phones before entering the lab.
7. Do not bring sport equipments or bulky objects with you to the lab. They take up extra space and may hinder the evacuation process.

Working in the Lab

1. Follow your TA's instruction on wearing goggles, gloves, lab coats, and other protective gear. **Wear goggles** to protect your eyes. When instructed, wear them continuously. You will be ejected from the lab and receive a zero if the instruction to wear goggles is not followed. Do not wear contact lenses to the lab; in case of a chemical spill, contact lens may make it impossible to flush your eyes.
2. Burns are the most common injuries in chemistry labs. Burns are painful, can be serious, and may take a long time to heal. Avoid touching hot objects directly with your bare hands. Learn to use tongs properly by practicing with a cool crucible; be sure to always keep one hand behind your back to resist the automatic reaction to reach out to save a falling object. Remember that heated items (burners, rings and ring stands, hotplates, etc) may take 10 to 30 minutes to cool down. And always turn off the burners when not in use.

3. Inhaling chemical fumes may cause serious medical conditions. Avoid breathing fumes by holding the test tube away from your face. Never put a test tube directly under your nose. If you need to detect the presence of an odor, hold the test tube about a foot away from your face and use your hand to gently waft some of the fumes toward your nose.
4. Always inspect the glassware to ensure it is not damaged. If a glassware is cracked, chipped or broken, report to your TA immediately. Report all accidents even if they seem trivial or may be embarrassing.
5. Never work in the laboratory without a TA present.
6. Never attempt unassigned experiments on your own. Discuss first with your TA in case there was a change in either the text or verbal instructions of the assigned experiment.
7. Notify the TA immediately if you witness dangerous behaviors by your neighboring lab mates. Watch out for each other and remind each other to be careful when needed.
8. Do not toy around with or abuse the lab equipments.

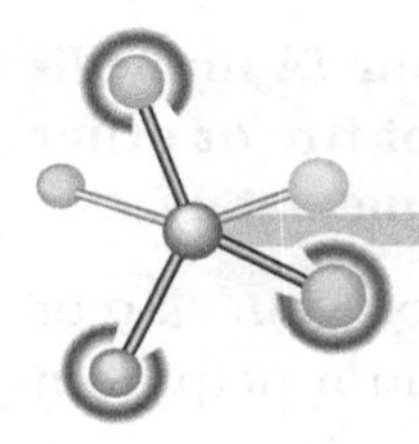

Behavior

The following "misbehaviors" have been observed before. They are strictly prohibited in this laboratory.

1. Do not stand on the stool to reach anything or to pour chemicals into an apparatus. Ask your TA for assistance, there is a safe way that eliminates any need to stand on the stool.
2. Do not sit on the bench even when there is no chemical on the bench.
3. Do not bring food into the laboratory even if you do not intend to eat in the laboratory (You may hide it in your bag, but we still do not recommend it).
4. Do not sit on the floor next to the bench that has something boiling on it.
5. Do not sit on the floor when there is nothing boiling on the bench.
6. Just do not sit on the floor.
7. Do not joke in a hysterical manner, nor in a continuous manner, with your partner or anyone in the laboratory. This behavior raises safety concerns and distracts oneself and others from learning and finishing the work.

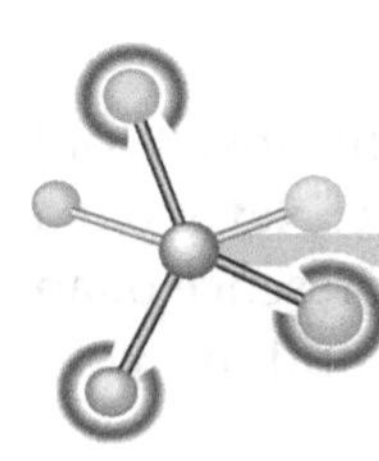

Post-Lab: Waste Disposal and Cleanup

1. Follow the TA's guidelines on disposal of chemical wastes. Containers will be provided for specific chemical materials.
2. Put broken glass in the proper container in the front of the lab. **Never put glass in the common trash bin.**
3. Clean up after yourself to keep the lab space and equipment clean.
4. Wash your hands with soap and water when you are done with each experiment and when you exit the lab. Do not touch your face without first washing your hands.
5. Do not take any chemical outside the lab.

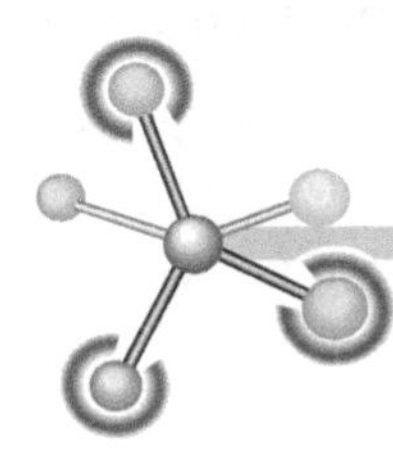

In Case of an Accident

Stay calm and get help. Report any injury or accident to your TA immediately, no matter how minor you think they are. In case of fire or other emergency, follow evacuation procedures and evacuate immediately.

1. For chemical spills: If the chemical gets in your eyes or on your skin or clothing, stop what you are doing immediately and call for help from the TA.
   a. If a **chemical splashes in your eyes**, your TA will help you to the eyewash station. Quickly drench your face and flush your eyes thoroughly for approximately 10 to 20 minutes. Keep your eyes open and rotate your eyeballs to be sure water reaches everywhere.
   b. If a **chemical splashes on your skin or clothing**, stay where you are and notify your TA. If the spill affects a large area of your skin, you will be taken quickly to the safety shower. Remove all clothing, shoes, and jewelry that may be contaminated and flood the affected area with cold water.
   c. If there is a **chemical spill** on the bench, floor, reagent shelf or anywhere in the lab, notify your TA and neighbors immediately. If there is a potential for the spill to explode or burn, call for everyone to evacuate.
2. For cut and scrapes, report to your TA no matter how minor the cut. If there is **serious bleeding**, apply direct pressure with a clean, preferably sterile dressing. Have someone immediately call 711. Students should avoid contact with another person's blood and should stay clear of the area until told that it is safe to return.

lab 1

Volume of Mixing and Liquid Crystals

There is a lot of space between molecules in a gas phase; there is also space between molecules in a liquid or a solid although a person may not "feel" it. This space between molecules is called *interstitial space,* and is governed by the intermolecular non-covalent interactions and the thermal energy of the molecules. The intermolecular interactions (ionic bonding, ion-dipole interactions, dipole-dipole interactions, induced dipole-dipole interactions, hydrogen bonding and van der Waals dispersion forces) and the thermal energy of molecules also determine the phase of the molecules (solid, liquid, gas or liquid crystal), as well as whether two different liquids are miscible or not.

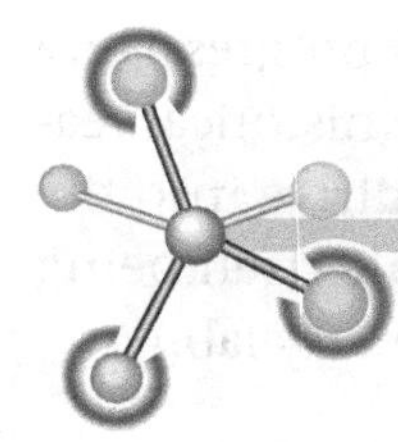

Partial Volume

When two liquids mix, depending on the chemical nature of the liquids, the mixed volume may or may not be equal to the sum of the original two volumes. Often, the mixed volume is not the sum of the two volumes. However, the volume of mixing, V_{mix}, is given by equation 1 below.

$$V_{mix} = n_1V_1 + n_2V_2, \ldots \qquad \text{(Eq. 1)}$$

where n_1 and n_2 are the number of moles of each of the liquid and V_1 and V_2 are called the partial molar volume. The partial molar volume of a component (V_1, V_2, . . .) is predetermined and *is equal to the increase in the volume of solution when one mole of the component is mixed in a large volume of specified composition.* For example, to find out the partial volume of water in a solution of 70 mol % water and 30 mol % ethanol, scientists will add 1 mole of water to a very large volume of solution containing the 70 mol % of water and 30 mol % of ethanol, and then measure the increase in volume. A "very large volume" implies at least about 100 times the volume of one mole of the added component.

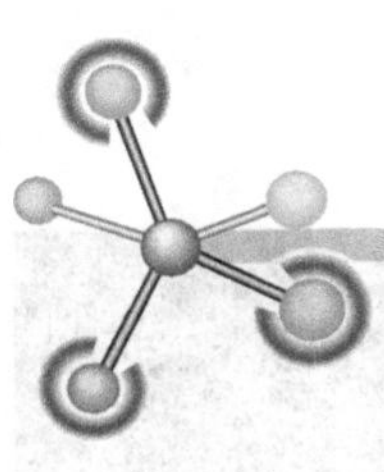

Liquid Crystals

Liquid crystal is a state of matter that exists between solid and liquid. Molecules in the liquid crystal state have orientational order, but no positional order of molecules (Scheme 1), so they can move or flow in liquid crystal with a preferred orientation. As a consequence, liquid crystals have the optical properties of a crystal such as a diamond but the fluidic property of a liquid. Because of the fluidic property, the orientation of the molecules in the liquid crystal phase can be manipulated by external stimulates, such as temperature and electric field. Controlling the orientation controls the optical properties, and thus creates a wide range of applications include display of television and computer, thermal sensor (i.e., thermometer), switching a window from transparent and not transparent, and others.

Scheme 1

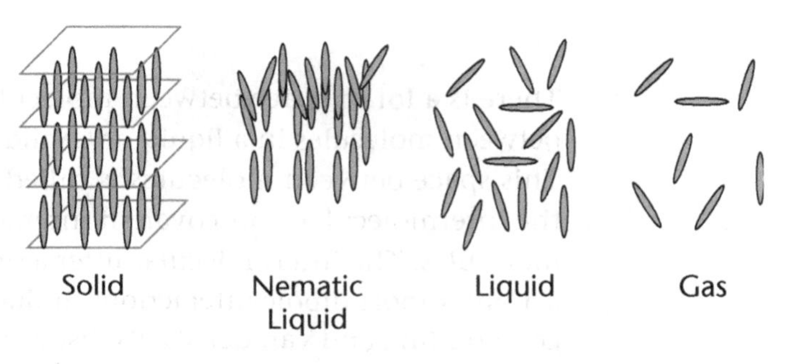

Sometimes a liquid crystal can have layers of molecules with progressively twisted orientation. This is called cholesteric liquid crystal (for historical reasons) (Scheme 2). This phase is caused by having a chiral additive into the normal liquid crystal or by having the liquid crystal molecule itself being chiral. ("Chiral" or chirality is discussed in the upcoming symmetry lab.)

Scheme 2

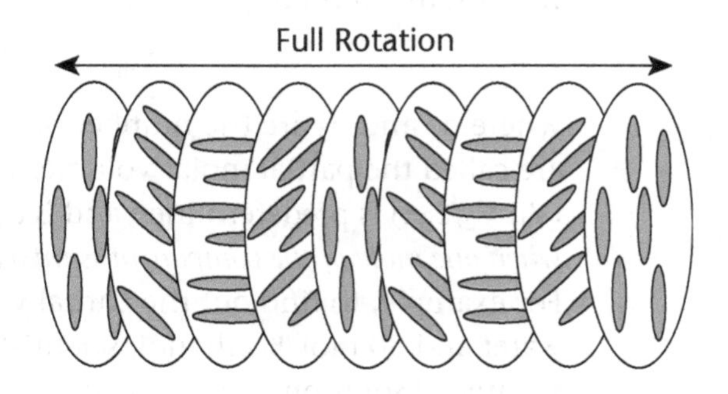

This phase of the molecules often exhibits beautiful colors. The angle of the twist of each ordered layer of molecules, and thus the distance of a full turn of the oriented molecules, is extremely sensitive to the temperature. As the temperature increases or decreases, the full turn will increase or decrease, respectively. This change in molecular organization is accompanied with a change in the color of the materials.

Liquid crystals change the polarization of light. Lights are waves. Normal lights have the waves oscillating in all directions perpendicular to the light's traveling direction. However, some waves in oscillating directions can be blocked (or filtered) more than other directions by highly ordered material, such as crystals and liquid crystals. This phenomenon is due the light being absorbed more by one axis of the ordered materials than the other axes. Polarizers are made of highly ordered materials. This materials will block out most of the light except for light wave that oscillates parallel to long axis of the order of the materials in the polarizers. Thus, when two such polarizers are stacked together with the orders in the materials 90° relative to each other, ALL lights will be blocked out. Looking through such crossed polarizers, it would appear to be black and non-transparent. However, when one of the polarizers is rotated away from 90°, light will start to leak through again, and one can then look through the polarizers. When the two polarizers are stacked with the order in the materials parallel to each other, maximum amount of light will pass through.

When liquid crystals (or crystals) are placed in between the polarizers, plane polarized light passing through the first polarizer[1] will be polarized again by the liquid crystals to another angle, and thus will leak out through the second polarizer.[2]

1. First and second polarizers are arbitrary, depending on the location of the light source.
2. In the very rare case, in which the molecular order in the liquid crystal aligns parallel to the axis in one of the polarizers, then light will not leak out. In such a case, one just needs to rotate the liquid crystal between the polarizers, then light will leak out again.

$$\# \text{moles} = \frac{\text{density} \times \text{volume}}{\text{molecular weight}}$$

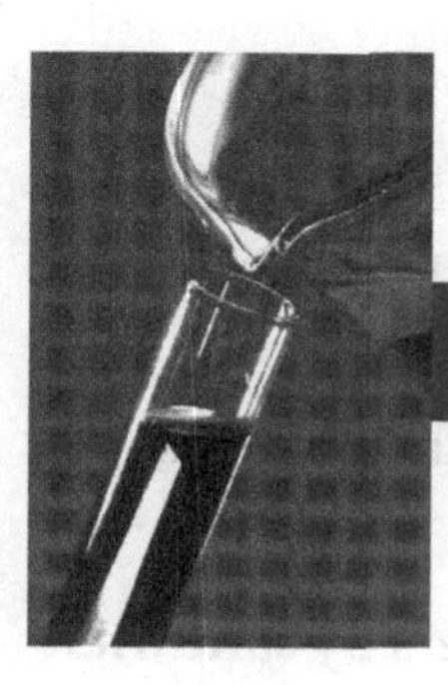

Lab 1 Exercise

PURPOSE To experience interstitial nature of chemicals by measuring the volume of mixing and the color change as a cholesteric liquid crystal is subject to temperature change.

PROCEDURE ❶ **Determination of combined volumes of two liquids.**

1. Measure the appropriate amounts (Table 1) of ethanol and water in two separate graduated cylinders.
2. Transfer the ethanol to the graduated cylinder containing the water and record the resulting volume after mixing (Table 1).
3. Repeat procedures 1 and 2 for the four quantities (runs) given in Table 1.

table 1

Data for Mixing of Ethanol and Water

	RUN 1	RUN 2	RUN 3	RUN 4
Ethanol (mL)	10	15	5	20
Water (mL)	10	5	15	6.15
Mixed volume (mL)	19.2	19.0	20.0	24.9

PROCEDURE ❷ **Making cholesteric liquid crystals.**

1. Two samples of cholesteric liquid crystals, sample "Low" and sample "High," will be prepared. For each sample, mix chemicals A, B, and C in a vial (amounts are indicated in Table 3 in Post-Lab Questions). The mass of each chemical is arbitrary as long as it is within the targeted mass range shown in the table; however, you need to record the *exact mass* you have weighed.
2. To thoroughly mix the compounds in each sample, warm each sample in a hot water bath until the mixture melts in the vials.
3. Let the vials cool to room temperature for 10 minutes or so. Record any visual changes that occur for each sample as it cools.
4. Put the vials into a cold water bath for 10 seconds. Record any physical changes that occur.

5. Place each vial between your hands and roll back and forth for 45 seconds. Inspect and record visual changes.
6. Allow the vial to return to room temperature and record the final changes.

PROCEDURE 3 **Observing light passing through a pair of crossed polarizers by sandwiching the liquid crystals in between.**

1. Scoop some of the liquid crystal sample you prepared, and sandwich it between two glass slides.
2. Place the two glass slides sandwiching the liquid crystal between two polarizers. Note your observations. Liquid crystal changed polarization of light and allowed light to leak through.
3. Rotate one of the polarizers (the top one) until the area outside the liquid crystal is of maximum darkness. Note your observations.

Light still passes through

CALCULATIONS

1. Calculate the number of moles for each of the samples in Table 1. Enter your results on Table 1.

   Given: The density of ethanol is 0.785 g/cm^3; density of water is 0.997 g/cm^3.

   NOTE: The TA will teach the students on the calculation.

2. Given a mole fraction of 0.5, the partial volume of ethanol is 57.4 mL and the partial volume of water is 16.9 mL: [15 pt.]
   a. Calculate the theoretical volume of mixing at 0.5 mole fraction of ethanol and water. [HINT: Use Eq. (1).]
   b. Compare your experimental results (i.e., the mixed volume as recorded in Table 1) and compute a percentage of error.

$$\% \text{ error} = \frac{\text{experimental value} - \text{theoretical value}}{\text{theoretical value}} \times 100\% \quad \text{(Eq. 2)}$$

STUDENT NAME	Brandon Breazeale
PARTNER	Josh Brown / Harrison Arciprete
TA NAME	Lu Yu
DATE	2/5/13

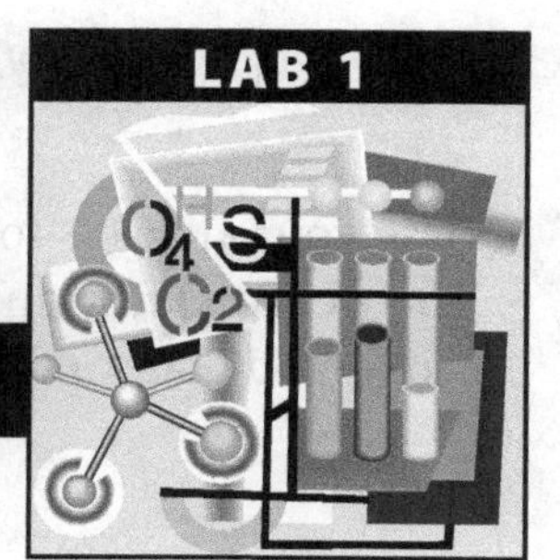

POST-LAB QUESTIONS

Lu Yu

02/04/2013

1. Sometimes a student's liquid crystal sample appears to respond to temperature differently from others. Provide a possible explanation. [5 pt.]

2. Draw the molecular structure of water showing where the location of the lone pairs of electrons relative to the O—H bonds. Name the most important molecular interaction for liquid water. [10 pt.]

3. It is very clear that the gas phase of a substance has a lower density than its liquid phase. In general, then, which has a higher density: the solid state of a substance or its liquid state? [5 pt.]

4. Which has a higher density, water or ice? [5 pt.]

5. Use your answers for Question 2 and create schematic drawings to provide an explanation for your answer to Question 4. [HINT: Loosely speaking, ice has a diamond structure and liquid water has a collapsed diamond structure.] [20 pt.]

6. Draw a scheme showing possible molecular interactions for liquid water and ethanol. [10 pt.]

7. Describe briefly what you see when you put the polarizer sandwiched liquid crystal sample on top of the light source.

10pts each

8. When sandwiched between a pair of crossed polarizers, cholesteric liquid crystals always cause light to leak out, but other crystals can always have an orientation that blocks out light. Why?

Data and Calculations

table 2

Calculations for Mixing of Ethanol and Water [20 pt.]

	RUN 1	RUN 2	RUN 3	RUN 4
Mixed volume (mL)	19.2 mL	19.0 mL	20.0 mL	24.9 mL
# of moles of ethanol				
# of moles of water				
Mole ratio				

Calculations

1. Theoretical volume of 0.5 mole fraction of ethanol and water. [10 pt.]

2. Percentage of error. [5 pt.]

table 3

Data and Observations for Liquid Crystals [10 pt.]

SAMPLE LOW	A (Ch. oleylcarbonate)	B (Ch. nonanoate)	F (Ch. propionate)
Targeted mass range	95–105 mg	195–205 mg	33–37 mg
Actual mass	105mg	202g	33mg
Observations during cooling	Low is turning purple, then back to milky		
Observations after 10 sec in cold bath	Turned light blue clear, shiny		
Observations after 45 sec in hands	Opaque clear, much less purple		
Observations during final cooling	almost completely white		
SAMPLE HIGH	**(Ch. oleylcarbonate)**	**(Ch. nonanoate)**	**(Ch. propionate)**
Targeted mass range	145–155 mg	195–205 mg	47–53 mg
Actual mass	147mg	201mg	49 mg
Observations during cooling	High is turning a milky clear, with a small amount of purple		
Observations after 10 sec in cold bath	Turned dark purple with white edges		
Observations after 45 sec in hands	Opaque clear, much less purple		
Observations during final cooling	unchanged, maybe slightly more purple		

Last question: Describe one concept that you understand better as a result of this lab. Be specific. (10 pt.)

lab 2

Enthalpy and Entropy of Dissolving Salts

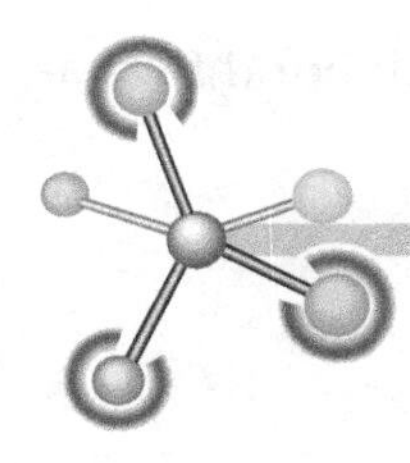

Enthalpy

The energy or enthalpy change associated with the process of a solute dissolving in a solvent is called the *heat of solution* (ΔH_{soln}). In the case of an ionic compound dissolving in water, the overall energy change is the net result of two processes—the energy required to break the attractive forces (ionic bonds) between the ions in the crystal lattice, and the energy released when the free ions form dipole-ion attractive forces with the water molecules.

Heat of solution and other enthalpy changes are generally measured in an insulated vessel called a *calorimeter* that reduces heat loss to the atmosphere outside the reaction vessel. The process of a solute dissolving in water may either release heat into the aqueous solution or absorb heat from the solution, but the amount of heat exchanged between the calorimeter and the surroundings should be minimal. When using a calorimeter, the reagents being studied are mixed directly in the calorimeter and the temperature is recorded before and after the reaction has occurred. The amount of heat change occurring in the calorimeter may be calculated using the following equation:

$$q = m \times c \times \Delta T \qquad \text{(Eq. 1)}$$

where m is the total mass of the solution (solute plus solvent), c is the specific heat of the solution, and ΔT is the observed temperature change. The specific heat of the solution is generally assumed to be the same as that of water, namely, 4.184 J/(g°C).

When a salt dissolves in water, the ions are surrounded by water molecules. This process involves breaking the solute-solute (cation-anion) interactions and forming the solute-solvent interactions (ion-water). For any dissolving process, the solvent-solute interactions (attraction) can be larger or smaller than the combined solute-solute and solvent-solvent interactions. To determine the sign of enthalpy of a reaction or a process, Table 1 shows a generally true guideline.

table 1

Determining the Sign of Enthalpy

RELATIVE STRENGTHS OF MOLECULAR INTERACTIONS	ENTHALPY OF DISSOLVING	SIGN OF $\Delta H_{solvation}$
Solute-solvent > solute-solute and solvent-solvent	Exothermic	(–)
Solute-solvent < solute-solute and solvent-solvent	Endothermic	(+)

This table can be rationalized by the principle that, in general, bond formation releases heat, whereas breaking bonds absorbs heat.

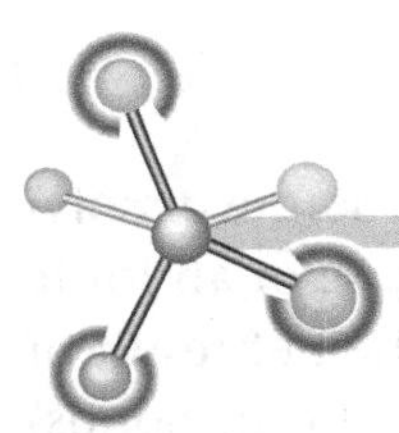

Free Energy

The spontaneity of any reaction is governed by two thermodynamic concepts, enthalpy, and entropy. *Entropy* is a measure of the degree of disorderliness of a system. Although all spontaneous processes produce an increase in the entropy of the universe, increasing the entropy of a particular system does not guarantee spontaneity. Likewise, while exothermic reactions tend to be spontaneous, they are not always, as is the case of water freezing at 5°C.

Gibb's free energy calculation is used to predict whether a process will be spontaneous or non-spontaneous. Gibb's free energy, G, is defined as

$$\Delta G = \Delta H - T\Delta S \quad \text{(Eq. 2)}$$

where T is the absolute temperature (and should be constant), and ΔH and ΔS are the enthalpy and entropy changes, respectively, for the given process. For a reaction to be spontaneous, ΔG needs to be negative.

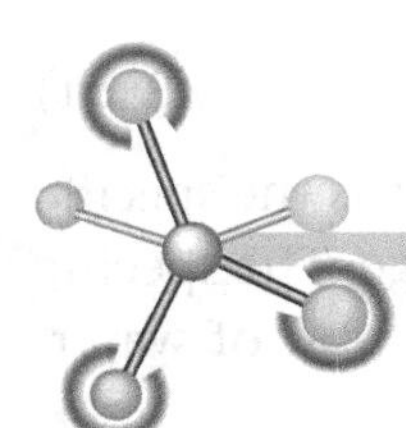

Science on Solvation of Salts

When different kinds of salts dissolve in water, sometimes heat is released resulting in an increase in the temperature of the solution; sometimes heat is absorbed leading to a decrease in the temperature. For a long time (nearly 200 years), people do not understand why this is so, and thus the science remains empirical. However, after numerous studies on different kinds of salts, chemists recently discovered a pattern, from which they developed the following new concept.[1] When salts dissolve in water, not every kind of salts will separate into cations and anions with individual ions surrounded by water molecules. Some salts will have their cations and anions close to each other in pairs; and each pair is solvated in water. The science is related to the sizes of the ions, and is organized into two cases described below.

First, salts made of small anions and big cations, or big anions and small cations, like cesium fluoride (CsF) or lithium iodide (LiI) produces hot solutions when dissolved in water. This phenomenon is understood as large number of hydrogen bonds being formed between the completely solvated ions and the water molecules. The strong interaction of the solvated individual ions with the water molecules results in more hydrogen bonds formed between ions and water molecule in comparison to the number of hydrogen bonds broken between the water molecules. Thus, on dissolving salts made form small cation and large anion, or large cation and small anion, the overall process is bond making, which releases heat. The process is exothermic and hot solutions are produced for dissolving these salts in water.

Second, people observe that salts made of small anions and small cations like sodium fluoride (NaF), or big anion and big cations, like cesium iodide (CsI) produces cold solutions when dissolved in water. This phenomenon is understood as the tendency of ions to exist as neutralized ion-pairs. On dissolving the salts in water, only limited number of hydrogen bonds are formed between the water molecules and neutralized ion pairs. The weak interaction of the neutralized ion-pairs with the water molecules results in less number of bonds formed between the two in comparison to the number of hydrogen bonds that is broken between the water molecules. Thus, on dissolving salts that consists of small cation and small anion or (big cation and big anion) the overall process is bond breaking, which absorbs heat. The process is endothermic, and such salts produce cold solutions when dissolved in water.

These different solvation of salts is pictorially represented in (Fig. 1).

For a detailed understanding of the above phenomena, we consider that salts in water act as neutral species made of positive and negative point charges wherein the point charges are located at the center of the ionic radii.[2] First, let's consider salts with mismatched sizes, like large cation and small anion cesium fluoride (CsF) or small cation and large anion lithium iodide (LiI) in water. The smaller cation or anion of the salts can strongly interact with water, small anion F^- and small cation Li^+ can more readily interact with medium sized water molecules in comparison to the large sized oppositely charged counterpart (Cs^+ or I^-). Such strong interaction solvates only small ions, and thus separates them from the large ions. Hence, when salts like CsF or LiI are dissolved in water it produces hot solution. The process is exothermic because small ions strongly interact with water and are completely solvated in the aqueous solution.

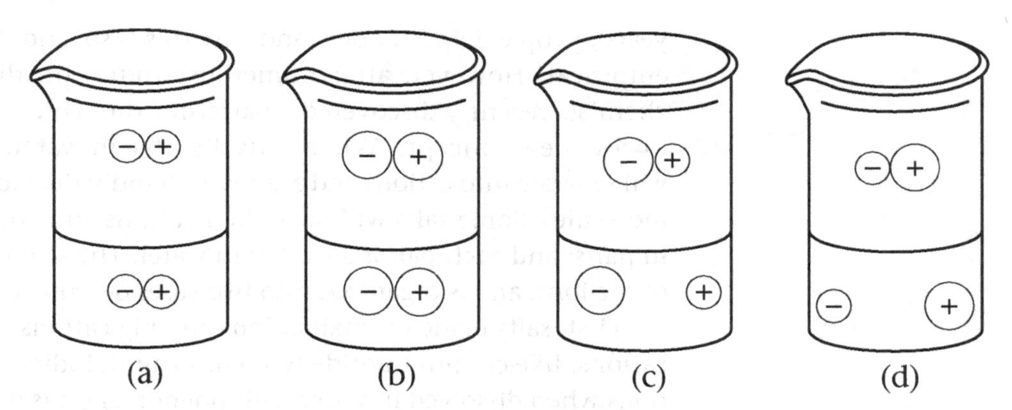

FIGURE 1. Behavior of salts in water. Salts with matched sizes like (a) sodium fluoride (NaF), and (b) cesium iodide (CsI) remain as neutralized ion-pair in water. Salts with mismatched sizes like (c) lithium iodide (LiI), and (d) cesium fluoride (CsF) are completely solvated as individual ions in water.

On the other hand, when salt formed by small cations and small anions like sodium fluoride (NaF) is dissolved in water, the small cation Na^+ with positive charge at the center can get closer to the center of the small negative charge of anion F^- more easily than it can get closer to the partial negative charge of oxygen in water molecule. In contrast, for a salt made of a large cation and large anion like cesium iodied (CsI), the medium sized water molecule have higher tendency to stay closer to other water molecules rather than the large cations or anions because the charge distance between two water molecules is smaller than between a water molecule and a bigger ion. Hence, when salts like NaF or CsI are dissolved in water it produces cold solution. The process is endothermic because these salts do not interact with water as strongly and are not completely solvated in the aqueous solution but instead remain as neutralized ion pairs in aqueous solutions.

In summary, enthalpy for dissolving ionic salts in water is dependent on the hydrogen bonds formed between water and interacting ions and the ionic sizes of ions that form the salts. The process of dissolution in water will be exothermic for salts formed by mismatched ion pairs, and endothermic for salts formed by ion pairs of the same size.

In the following experiment, the above science will be tested by dissolving different kind of salts in water and measure change in temperature for the dissolving salts. The size of cations and anions are documented in Table 2. Water, H_2O, is being thought as a molecule with a positive and negative charge (they call such a molecule zwitterion) with a radius of 106 pm for the cationic portion and 178 pm for the anionic portion.

table 2

Ionic Radii of Different Ions in Picometer (pm)[a]

SIZE	CATIONS	RADII (PM)	ANIONS	RADII (PM)
Small	Li	76	F	133
	Na	102	—	—
Big	K	138	Cl	181
	Cs	167	I	220

[a] 1 pm = 10^{-12} m; 1 m = 10^{12} pm = 10^{9} nm; 1 nm = 10 Å; 1 Å = 100 pm.

Endnotes

1. Collins, K. D., Charge density-dependent strength of hydration and biological structure. *Biophysical Journal* **1997**, 72, (1), 65–76.
2. Collins, K. D., Ions from the Hofmeister series and osmolytes: effects on proteins in solution and in the crystallization process. *Methods* (San Diego, CA, United States) **2004**, 34, (3), 300–311.

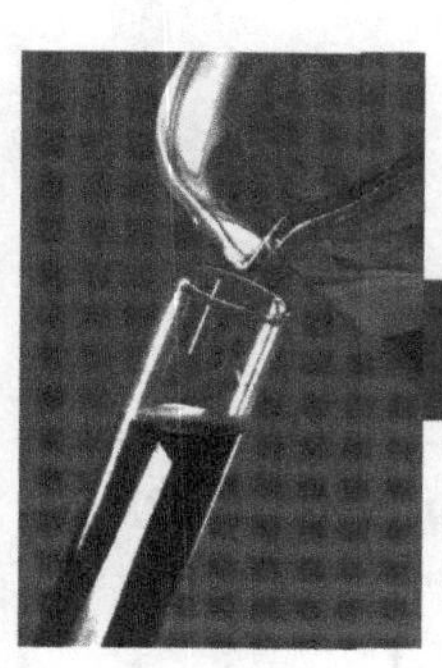

LAB 2 EXERCISE

PURPOSE To determine the enthalpy change that occurs when an ionic salt dissolves in water.

PROCEDURE ❶

1. Place a small amount of each of the salts listed in Table 3 on a dry watch glass. They may all be placed on the same watch glass but be sure they do not mix. Allow the watch glass to sit undisturbed on the bench until the end of the lab.
2. Weigh and record the mass of an empty, dry styrofoam cup.
3. Tare the balance and add approximately 1.5 g of potassium chloride to the cup. Record the actual mass of potassium chloride in Table 5 in Post-Lab Questions.
4. Measure the appropriate volume of water (H_2O) in a graduated cylinder as specified in Table 3.
5. Record the temperature of water inside the cylinder as the initial temperature of the solution, T_i, in Table 5 in Post-Lab Questions. Make sure that the bulb of the thermometer is fully submerged in the water.

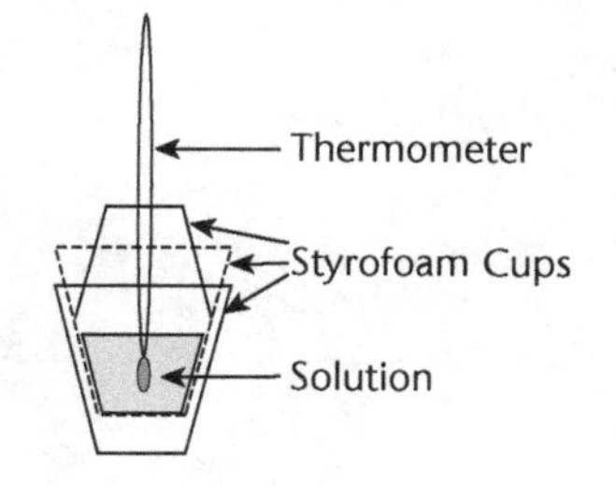

6. Transfer the water from the graduated cylinder into the cup. Cover the solution immediately with an inverted cup.
7. Use the thermometer as a stirrer to make sure that KCl completely dissolves in the water. Record the temperature when the solid has completely dissolved as final temperature, T_f, in Table 5 in Post-Lab Questions.
8. Measure and record the final mass of the solution and cup.
9. Follow the same procedure for Runs 2 through 5. Use the salt, vessel, and amount of water as specified in Table 3.
10. When all data have been recorded, re-examine the salt samples placed on the watch glass in Step 1. Have any of the salts obviously absorbed moisture from the atmosphere? Record your observations of the appearance and change of the crystals.

table 3

Solutions

	RUN 1	RUN 2	RUN 3	RUN 4	RUN 5
Salt	KCl	KCl	NaCl	LiCl	LiCl
Mass of salt (g)	1.5	3.0	1.5	1.5	1.5
Container	coffee cup	coffee cup	coffee cup	coffee cup	glass beaker
Volume of water (mL)	25	10	25	25	25

CALCULATIONS For each run, calculate the following and record in Table 6:

1. Calculate the mass of the final solution and the ΔT for the dissolving process.

2. Assuming that no heat escapes from the vessel, compute the heat of the solution, $q_{solution}$, using Eq. (3) below:

$$q_{solution} = -q_{vessel} = -(m_{solution})(C_{water})(\Delta T)$$
$$= -(m_{salt} + m_{water})(C_{water})(\Delta T) \quad \text{(Eq. 3)}$$

where mass is in gram, and C_{water} is the specific heat of water, which is 4.184 J/(g°C).

3. Compute the molar enthalpy for the dissolution of the salt in water, ΔH_{soln}, using Eq. (4) below:

$$\Delta H_{soln} = \frac{\Delta q_{soln}}{n_{salt}} = \frac{-q_{soln}}{n_{salt}} \quad \text{(Eq. 4)}$$

where n_{salt} = moles of KCl, NaCl or LiCl
FW_{KCl} = 74.54 g/mol
FW_{NaCl} = 58.45 g/mol
FW_{LiCl} = 42.39 g/mol

4. Compute for the percent error (% error) for ΔH_{soln} using Eq. (5) below. For your accepted values, use those values calculated in pre-lab Question 4:

$$\%\ error = \frac{\text{experimental } \Delta H_{soln} - \text{theoretical } \Delta H_{soln}}{\text{theoretical } \Delta H_{soln}} \times 100\% \quad \text{(Eq. 5)}$$

PURPOSE To measure the heat of solvation of different ionic salts.

PROCEDURE 2

1. Weigh 0.76 g of cesium fluoride listed in Table 4 into an empty styrofoam cup.
2. Measure 10 mL of water in a graduated cylinder. Record the temperature of water inside the cylinder as the initial temperature, T_i, in Table 7 in Post-Lab Questions. Make sure that the bulb of the thermometer is fully submerged in the water.
3. Transfer water from the graduated cylinder into the styrofoam cup containing weighed cesium fluoride. Cover the solution immediately with another inverted styrofoam cup. (See accompanying Figure in Procedure 1; do not throw this solution. You will be using this solution in Procedure 3.)
4. Use a thermometer to stir the mixture until cesium fluoride has completely dissolved in water. When the solid completely dissolves, record the temperature of the solution as the final temperature, T_f, in Table 7 in Post-Lab Questions.
5. Repeat steps 1 to 4 using different salts each time (amount of salt to be weighed each time is different and is listed in Table 4) and record the final temperature for each salt in Table 7 in Post-Lab Questions.

table 4

The Amount of Salt Needed in 10 mL of Water to Make a 500 mM Solution

SALTS	CsF	LiI	KCl	LiF	KI
Salt mass (gram)	0.76	0.67	0.37	0.13	0.67

PURPOSE

To study the enthalpy change when two ionic salts formed by mismatched sizes of cations and anions are dissolved in water.

PROCEDURE 3

1. Measure the temperature of CsF and LiI solutions (prepared in Procedure 2) again and record in Table 8 in Post-Lab Questions. (Wait for both the solutions to achieve approximately same the temperature as of water at room temperature.)
2. Transfer the CsF solution (from Procedure 2) into the styrofoam cup containing lithium iodide solution (from Procedure 2). Cover the mixture immediately with another inverted styrofoam cup and record the temperature using a thermometer as the final temperature, T_f, in Table 8 in Post-Lab Questions.
3. Measure the temperature of the LiF and KI solutions and record in Table 8 in Post-Lab Questions (temperature of both the solutions should be approximately near the initial temperature of water T_i, if not wait for some time.)
4. Transfer the lithium fluoride solution (from Procedure 2) into the styrofoam cup containing potassium iodide solution (from Procedure 2). Cover the mixture immediately with another inverted styrofoam cup and measure and record the temperature using a thermometer as the final temperature, T_f, in Table 8 in Post-Lab Questions.

STUDENT NAME
PARTNER
TA NAME
DATE

LAB 2

POST-LAB QUESTIONS

1. Based on the experiment of dissolving lithium chloride, was the amount of heat absorbed in the breakup of the crystal lattice larger or smaller than the amount of heat released in hydrating the ions? Explain. [5 pt.]

2. If some water had vaporized as a result of an exothermic reaction, would your calculated heat of solution be too large or too small? State your reason in complete, concise sentences. [10 pt.]

3. A number of materials are "hygroscopic" or "deliquescent" (means they will absorb water from the atmosphere). Suppose your sample had absorbed water from the air before you weighed it. What effect would this have had on the on the value of ΔH_{soln}? Why? [HINT: This would have introduced two different errors.]

   Did any of the salt(s) used in this experiment exhibit deliquescence (absorb moisture from air and become wet)? What was the percentage of error for measuring the enthalpy for this salt? [15 pt.]

4. How does the temperature change when using the cups compare to the temperature change when using the beaker (without rubber stopper)? Why? [5 pt.]

5. Using the ΔG_f^0 values given below and the dissolution reaction written in pre-lab Question 4, calculate ΔG for the dissolving of KCl. [5 pt.]

$$\Delta G^0_{reation} = \sum nG^0_{f\,products} - \sum mG^0_{f\,reactants} \quad \text{(Eq. 6)}$$

where *n, m* is stoichiometric coefficients

Chemicals	ΔG_f^0 (kJ/mol)
$KCl_{(s)}$	–408.3
$K^+_{(aq)}$	–283.2
$Cl^-_{(aq)}$	–131.2

6. Using Gibb's free energy equation (T = 298 K), calculate the ΔS value for the dissolving of the potassium chloride. (Use the ΔG value calculated in the previous question and the ΔH value calculated in pre-lab Question 4.) [5 pt]

7. Predict and explain whether the dissolution of the following salts in water will be an exothermic or endothermic process? [5 pt]

   1. LiCl

   2. KCl

8. What algebraic sign appears on the entropy values? What does this mean? Why might this sign be expected? [5 pt.]

Data and Calculations [5pt]

table 5

Data

	RUN 1	RUN 2	RUN 3	RUN 4	RUN 5
Mass of vessel (g)					
Mass of salt (g)					
Mass of vessel + solution (g)					
T_i (°C)					
T_f (°C)					

Observations

	KCl	NaCl	LiCl
Appearance of salt at start of lab			
Appearance of salt at end of lab			

table 6

Calculations

	RUN 1	RUN 2	RUN 3	RUN 4	RUN 5
Mass of solution (g)					
ΔT					
q_{soln} (J) [5 pt.]					
ΔH_{soln} (kJ/mol) [5 pt.]					
[a]Theoretical ΔH_{soln} (kJ/mol)					
% error [5 pt.]					

[a] Use the data you calculated in the prelab.

Temperature of just water in the room: ________________ (T_i)

table 7

Observations for Procedure 2 [15 pt]

SALTS	SALT MASS (gram)	SOLUTION TEMP. (T_f)[a]	CLASSIFY IONS[b]	EXPECTED ΔH[c]	OBSERVED ΔH[c]
Cesium fluoride (CsF)					
Lithium iodide (LiI)					
Potassium chloride (KCl)					
Lithium fluoride (LiF)					
Potassium iodide (KI)					

[a] Temperature right after all the salt dissolves.
[b] Classify ion as big-big, small-small, big-small or small-big using Table 2.
[c] Write down positive or negative.

table 8

Observations for Procedure 3 [10 pt]

SOLUTIONS TO BE MIXED	TEMPERATURE BEFORE MIXING[a]		TEMPERATURE AFTER MIXING	EXPECTED ΔH[b]	OBSERVED ΔH[b]
CsF_{aq} and LiI_{aq}	CsF_{aq}	LiI_{aq}			
LiF_{aq} and KI_{aq}	LiF_{aq}	KI_{aq}			

[a] Please note that these solutions should have been cooled to room temperature. We record the temperature just for confirmation and for comparison to the temperature after mixing the solutions.
[b] Write down positive or negative.

Last question: Describe one concept that you understand better as a result of this lab.Be specific. (10 pt.)

STUDENT NAME
PARTNER
TA NAME
DATE

LAB 2

PRE-LAB ASSIGNMENT

1. Briefly explain the following terms: [30 pt.]
   a. Specific heat capacity of water
   b. Lattice energy
   c. Heat of hydration
   d. Hygroscopic
   e. Entropy
   f. Gibb's free energy

2. What information (data) is needed to calculate the enthalpy change for a reaction? [10 pt.]

3. Write the equation for the dissolution of potassium chloride in water. Repeat for sodium chloride. [10 pt.]

4. Using the information below, calculate the theoretical molar ΔH_{soln} for KCl, NaCl and LiCl. [30 pt.]

	ΔH:
$KCl_{(s)} \rightarrow K^+_{(g)} + Cl^-_{(g)}$	701 kJ/mol
$K^+_{(g)} \rightarrow K^+_{(aq)}$	–354.0 kJ/mol
$Cl^-_{(g)} \rightarrow Cl_{(aq)}$	–331.0 kJ/mol
$NaCl_{(s)} \rightarrow Na^+_{(g)} + Cl^-_{(g)}$	788 kJ/mol
$Na^+_{(g)} \rightarrow Na^+_{(aq)}$	–453.0 kJ/mol
$LiCl_{(s)} \rightarrow Li^+_{(g)} + Cl^-_{(g)}$	834 kJ/mol
$Li^+_{(g)} \rightarrow Li^+_{(aq)}$	–540.4 kJ/mol

Record these results in the theoretical ΔH_{soln} row of Table 6 of POSTLAB for later use.

5. In Table 4, why do we add different amount (by weight) of salts into the same amount of water? Can the results be compared? [5pt]

6. Look up the electronegativity values of oxygen and hydrogen. Draw the structure of the water molecule and indicate its polarity two ways: [10 pt.]

   a. Using δ+ and δ– to indicate the partial positive and negative charges on the atoms.

   b. Using an arrow to show the dipole moment of the water molecule (the crossed end of the arrow designates the positive end of the molecule).

7. Draw the arrangement of water molecules surrounding the hydrated lithium and chloride ions. Be sure to show the proper orientation of the water molecules. [5pt]

lab 3 Functional Groups of Organic Molecules: A Study of Polarity by Using Thin Layer Chromatography

Organic molecules are made of primarily carbon, hydrogen, oxygen, nitrogen, and a few other elements including sulfur and phosphorus. When only carbon and hydrogen are present, the molecules are called hydrocarbons (R). When other elements are present in the molecules, the structure of the organic molecules is considered to possess different functional groups, such as hydroxyl groups (ROH), ketone ($R_2C{=}O$), aldehyde (RHC=O), amine (RNH_2), carboxylic acids (RCOOH), amines, and others. These functional groups contribute to the overall polarity of a molecule. When the overall structure of an organic molecule is kept the same, but the functional group on the molecules are changed systematically at the same position on the molecule, it is often possible to compare the polarity of the functional group.

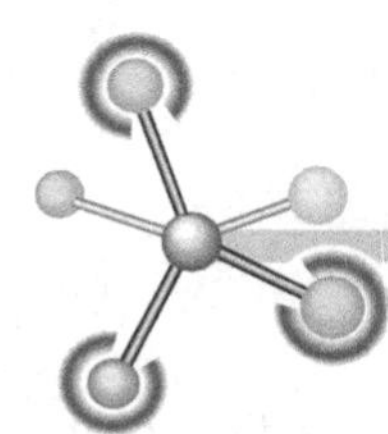

Principle of Thin Layer Chromotography (TLC).

The TLC plate is made of silica gel coated on a surface. Silica gel contains many hydroxyl groups and thus is a polar material coated on a surface. As such, the silica gel interacts (binds noncovalently through hydrogen bonds and dipole interactions, dispersion interactions) with polar molecules stronger than nonpolar molecules. When a solvent moves up the TLC plate due to capillary motion, the solvent will move (or drag) the nonpolar molecules faster than the polar molecules on the TLC plate because the interaction between the nonpolar molecules and the silica gel on the plate is weaker than the interaction between the polar molecules and the silica gel. In general, the relative polarity (i.e., which is more polar) of two molecules does NOT change when different solvents are used. But the molecules will be moved (or dragged) more by the polar solvent than the less polar solvent. This priniciple is true for all molecules (both polar and nonpolar). Polar solvents move nonpolar molecules faster than nonpolar solvents. This is because there are still weak interactions, such as van der Waals interactions and dispersion interactions, between the nonpolar molecules and the silica gel.

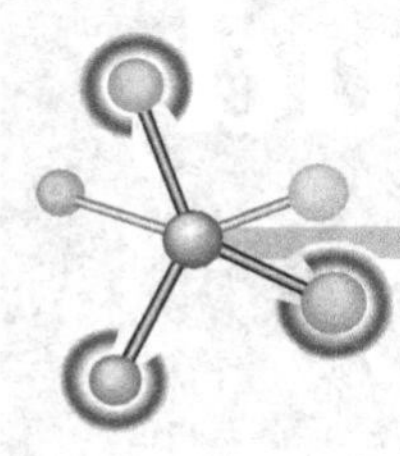

Running the TLC

The molecules or compounds of interest are first diluted in a volatile solvent, then spotted on a TLC plate using a capillary tube. (Note that the concentration of a TLC sample should never be too high.) The spots must be aligned on the same level marked by a pencil. The TLC plate is then put in a TLC chamber that contains some eluent solvent, which can be a mixture of two different organic solvents. The volume of the solvent must NOT touch the pencil mark of the spotted compounds. The TLC chamber is covered to avoid evaporation, which will disturb the capillary motion of solvents moving up the TLC plate, resulting in different results for the same experiments when they are run at different times or by different users.

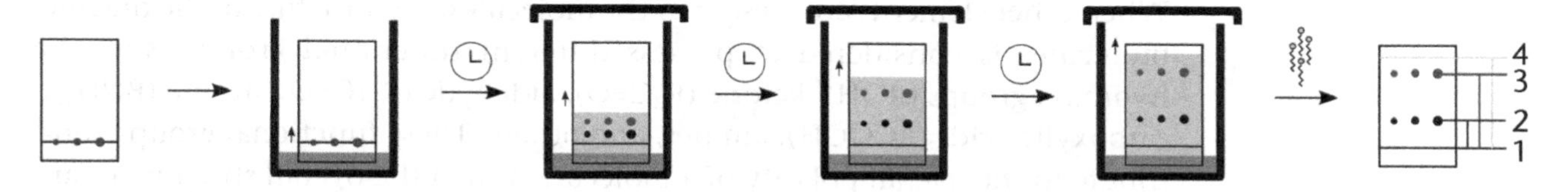

FIGURE 1. Running a thin layer chomotography (TLC). Released into the public domain by Quantockgoblin via Wikipedia.

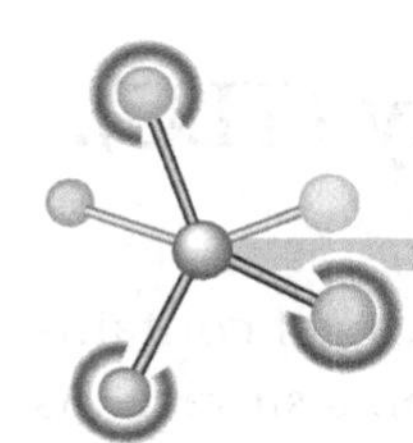

Rf (Retention Factor) Value.

One characterization of molecules is how far they will move on a TLC plate under the same conditions. The polar molecule will stay behind, and the nonpolar molecule will move faster and thus higher on the TLC plate. This statement is qualitative. Can one assign these distances with numbers so others can repeat the experiment with better precision? Yes, we use the Rf value to characterize these distances.

Rf = distance moved by the molecules/distance moved by the solvent.

For example Rf for the blue spot is (1_2)/(1_4); for the red spot is (1_3)/(1_4). We use this Rf value instead of just the distance from the baseline to the spot because everyone runs the TLC with an arbitrary length of time and thus the distance between the spot to the base will change from person to person, but the ratio (Rf) of the distance from baseline to the spot and the distance from the baseline to the solvent front does not change much.

Materials:
Eluent 1: Ethyl acetate/Hexane, 5/95 by volume.
Eluent 2: Ethyl acetate, 100%
TLC plate
Ferrocenecarboxaldehyde, (Dimethylaminomethyl) ferrocene,
Ferrocenecarboxylic acid, Ferrocenemethanol, Acetylferrocene, Ferrocene,

Retention Factor

$$Rf(A) = \frac{1-2}{1-4}$$

$$Rf(B) = \frac{1-3}{1-4}$$

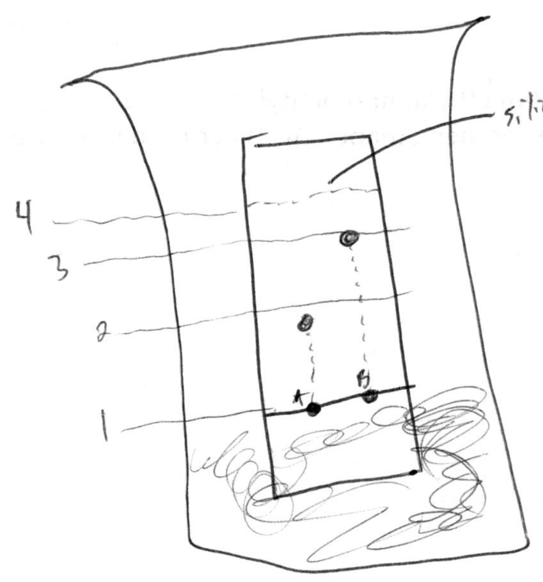

silica gel, polar material

B - nonpolar molecules
weak interaction with silica gel
moves fast

A - polar molecules
strong interaction with silica gel
moves slow

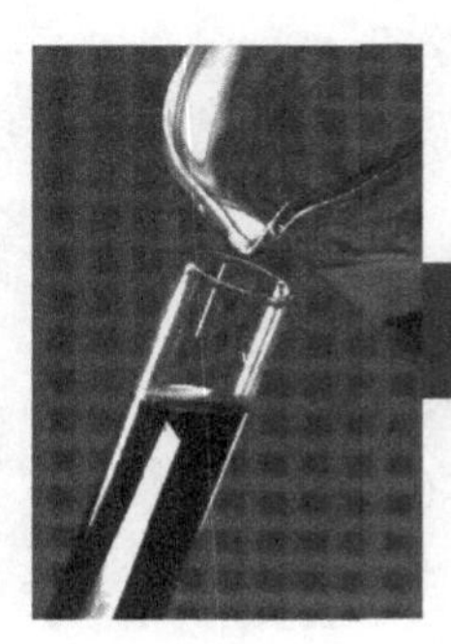

LAB 3 EXERCISE

Purpose: To understand the principle of thin layer chromatography and to study the polarity of organic function groups by using thin layer chromatography.

4 on a plate. Not 3

Procedure ❶: **Running the TLC with an unknown sample and known molecules side by side on the same TLC plate.**

1. Acquire a TLC plate. The TLC plate is composed of silica coated on a glass or plastic base. The properties of the silica and properties of the solvent are used to draw the spotted compounds up the plate at differing rates depending on the characteristics of the compound.
2. Draw a pencil line across the plate about 0.8 cm from the bottom of the plate.
3. Draw three small marks evenly spaced on the line. Beneath the marks label them A, U (for unknown), and B respectively, and in that order.
4. Dip a capillary in the sample A and then spot it on the first mark on the line.
5. Acquire another capillary and spot the unknown compound (U) on the middle mark.
6. Last, spot compound B on the right mark.
7. Place a piece of filter paper in the 50 ml beaker that will act as the TLC chamber.
8. Pour the solvent in the TLC chamber to a depth of about 0.5 cm.
9. Put the TLC plate into the chamber and the cover it with a watchglass. Make sure that the solvent does not touch the pencil mark.
10. Once the solvent front has reached near the top of the plate (say, about 0.5 cm), remove the watchglass, take the plate out, and immediately draw a line where the solvent front is. This is for calculating the Rf values.
11. Repeat the process on two new plates, keeping the unknown in the middle, but replacing compounds A and B with C and D, then after that repeat the procedures with E and F.

	E1 (6.0)	E2 (7.1)
A	5.4/6.0	.2/7.1
B	6.0/6	.7/7.1
C	.9/6	1.2/7.1
D	5.7/6	4.7/4.9
E	0	.8/4.9
F	.6/6.0	4.2/4.9
U	5.9/6	1.2/7.1

U 1.4/6.0 5.9/6.0 4.4/4.9 1.2/21

ab 4

Reaction Kinetics and Effect of Temperature

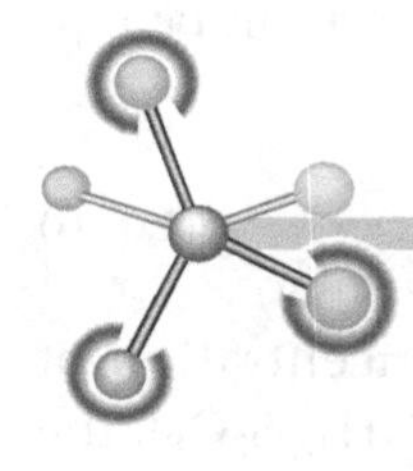

Reaction Rate and the Rate Law

The *reaction rate* refers to how fast a reaction proceeds or how fast the reactants are converted to products over time. *Chemical kinetics* is the study of reaction rates, how these rates are affected when the reaction conditions are varied, and the determination of a plausible mechanism on how the overall reaction occurred.

Consider a general reaction between substances A and B that yields products C and D and is represented by the equation:

$$aA + bB \rightarrow cC + dD \qquad \text{(Eq. 1)}$$

The rate of the reaction can be represented by the change in the disappearance of the reactants A ($-\Delta[A]/\Delta t$) and B ($-\Delta[B]/\Delta t$), and the appearance of the products C ($\Delta[C]/\Delta t$) and D ($\Delta[D]/\Delta t$) as a function of time. Considering the stoichiometric coefficients of the balanced chemical equation, the rate of reaction can be mathematically expressed in terms of the concentration of reactants A and B, and products C and D:

$$\text{rate} = -\frac{\Delta[A]}{a\Delta t} = -\frac{\Delta[B]}{b\Delta t} = -\frac{\Delta[C]}{c\Delta t} = -\frac{\Delta[D]}{d\Delta t} \qquad \text{(Eq. 2)}$$

where [A], [B], [C], [D] are concentrations, in molarity (mol/L), of reactants A and B, and products C and D, respectively.

At a given *constant temperature,* the rate law or rate equation is an expression of the rate of a particular reaction with respect to the concentrations of its reactants as shown below:

$$\text{reaction rate} = k[A]^x[B]^y \qquad \text{(Eq. 3)}$$

The exponents x and y are *determined experimentally* and represent the *order of reaction* with respect to the concentration of A and B, respectively; they are independent of the stoichiometric coefficients. The overall reaction order is the sum of the individual reaction orders. The *rate constant, k,* is a proportionality constant that relates the reaction rate with the concentrations of the reactants. A higher value of *k* denotes a faster reaction rate.

In this experiment, solutions of ammonium persulfate and potassium iodide proceed with the following reaction:

$$(NH_4)_2S_2O_8 + 3KI \rightarrow (NH_4)_2SO_4 + K_2SO_4 + KI + I_2 \qquad \text{(Eq. 4)}$$

The net ionic equation showing all the participating ions in the reaction is:

$$S_2O_8^{2-} + 3I^- \rightarrow 2SO_4^{2-} + I^- + I_2 \qquad \text{(Eq. 5)}$$

Because the rate of a reaction is a measure of the change in the concentration of the reactant (or the product) over time, the rate of reaction can be represented by:

$$\text{rate} = -\frac{\Delta[(NH_4)_2\,S_2O_8]}{\Delta t} \qquad \text{(Eq. 6)}$$

In this equation, $\Delta[(NH_4)_2S_2O_8]$ is the change in the concentration of $(NH_4)_2S_2O_8$ over the period of time, Δt. In order to measure $\Delta[(NH_4)_2S_2O_8]$, the reaction mixture is added with a little starch and a *specific* amount of $Na_2S_2O_3$ solution, which will quickly consume the I_2 immediately after it is generated:

$$2S_2O_3^{2-} + I^- + I_2 \;\; 2SO_4^{2-} + 3I^- \qquad \text{(Eq. 7)}$$

The experiment is purposely designed to have more molecules of $(NH_4)_2S_2O_8$ added than $Na_2S_2O_3$, so that when all of the $Na_2S_2O_3$ is used, the I_2 will appear again in solution. At this stage, the I_2 form a complex with starch to give a blue color.

$$I_2 + \text{starch} \rightarrow I_2{*}\text{starch complex (blue)} \qquad \text{(Eq. 8)}$$

From Eqs. (5) and (7), when one molecule of $S_2O_8^{2-}$ is consumed, two molecules of $S_2O_3^{2-}$ will be used to consumed in reacting with $I^- + I_2$. Consequently,

$$\Delta[(NH_4)_2\,S_2O_8] = \frac{1}{2}\Delta[Na_2S_2O_3] \qquad \text{(Eq. 9)}$$

$$\text{rate} = -\frac{\Delta[Na_2S_2O_3]}{2t} \qquad \text{(Eq. 10)}$$

Because we know the initial concentration of $Na_2S_2O_3$, and that it is completely consumed as blue color appears, $\Delta[Na_2S_2O_3]$ = initial concentration of

$Na_2S_2O_3$. Using Eq. (9), $\Delta[(NH_4)_2S_2O_8]$ can be calculated. Δt is measured by using a stopwatch to measure the time between mixing the reactants and appearance of the blue color.

The rate law can be written:

$$\text{rate} = k\,[S_2O_8^{2-}]^x[I^-]^y$$

where k is the rate constant and x and y are the orders of the reactants.

By varying the initial concentrations of $S_2O_8^{2-}$ and I^-, k, x and y can be experimentally determined. The rate constant can then be calculated.

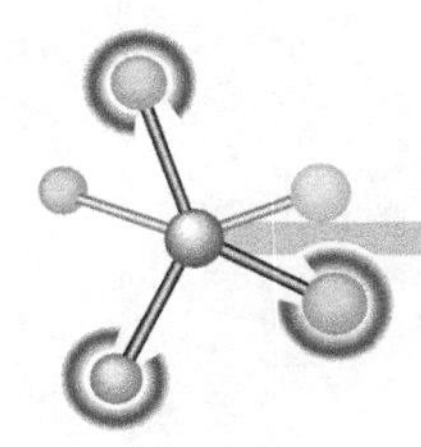

Temperature Effects on Reaction Rate

Temperature affects the reaction rate manifested by its effect on the rate constant, k. Based on the collision theory, the number of effective collisions between the reactant species in order to produce the products is proportional to the rate of reaction. For a particular reaction to proceed, a minimum collision energy—the *energy of activation,* E_a, must be surpassed by colliding reactant molecules. A reaction occurs when the reactants collide with sufficient energy. Thus, in general, the rate of a reaction increases with the frequency of reactant collisions that are above a certain threshold value of energy. This reaction rate–minimum collision energy relationship is represented by the *Arrhenius equation:*

$$k = Ae^{\frac{-E_a}{RT}} \qquad \text{(Eq. 11)}$$

where k = rate constant

E_a = activation energy

R = gas constant

A = frequency factor

Integration yields the equation:

$$\ln k = \frac{-E_a}{RT} + \ln A \qquad \text{(Eq. 12)}$$

Thus, if 1/T is plotted against ln k, the energy of activation can be obtained from the slope (slope = $-E_a/R$). So, what does increasing the temperature do to the energy of molecules and the frequency of collision, and thus rate of reaction? In this part of the experiment, Run 3 from Procedure 1 will be repeated at two different temperatures to observe the effect of temperature on reaction rate.

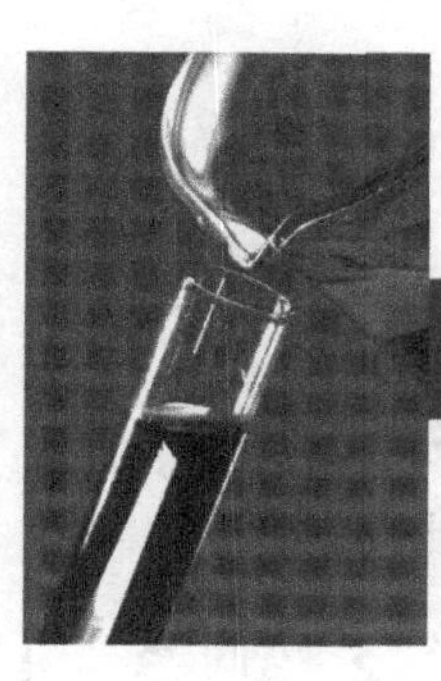

LAB 4 EXERCISE

PURPOSE To understand the effect of concentration of reactants and temperature on the rate of reaction.

PROCEDURE 1 **For each determination in Tables 2 and 3, repeat the following steps.**

1. Mix the appropriate volume of reagents as indicated in Table 1 (reagents A to E) in an Erlenmeyer flask. Place the flask on top of a white paper. This will allow you to observe changes in color of the solution as you monitor the progress of reaction.
2. Quickly add reagent F (0.16 *M* $(NH_4)_2S_2O_8$) to the beaker while swirling the solution. Start the timer immediately after the addition.
3. While stirring, observe the color change in the solution. When a blue color appears, stop the timer and write down the reaction time in Table 2A.
4. Complete a second trial of Run 1. If the reaction times are close, a third trial is not needed.
5. Repeat for Runs 2 and 3. Take the average of the two trials that agree most closely and average their times. This time will be used to calculate the reaction rate.

table 1

Reagents			
REAGENT	RUN 1	RUN 2	RUN 3
(A) 0.20 *M* KI	10.0 mL	10.0 mL	5.0 mL
(B) 0.2% starch	2.0 mL	2.0 mL	2.0 mL
(C) 0.01 *M* $Na_2S_2O_3$	4.0 mL	4.0 mL	4.0 mL
(D) 0.20 *M* KNO_3	—	—	5.0 mL
(E) 0.16 *M* $(NH_4)_2SO_4$	—	5.0 mL	—
(F) 0.16 *M* $(NH_4)_2S_2O_8$	10.0 mL	5.0 mL	10.0 mL

table 2A

Reaction Times (seconds)									
REACTION TIME, Δt	RUN 1 TRIAL			RUN 2 TRIAL			RUN 3 TRIAL		
	1	2	3	1	2	3	1	2	3
	9s	27s	35s	72s	60s		70s	74s	
	Avg.: 31s			Avg.: 66s			Avg.: 72.0s		

PROCEDURE 2

1. Using an ice bath, cool two beakers, each containing the mixture solution from Run 3, to 10°C *below* room temperature. Add the $(NH_4)_2S_2O_8$ to the flask containing reagents (A) through (E). Swirl in the ice bath to keep the reaction at 10°C below room temperature and record the reaction time in Table 3A.
2. Repeat the above procedure, using a hot plate to warm the two solutions to a temperature 10°C *above* room temperature.

table 3A

Reaction Times (seconds)			
RUN	3A	3B	3C
Temperature, T (°C)	13°C 10°C below	33°C 10°C above	23°C room temp.
Reaction time, Δt	2:12 2:03	:37 :35	1:10 1:11

CALCULATIONS Enter your results for calculations in runs 1 to 3 in Tables 2B and 3B.

1. Calculate the reaction rate for each trial using Eq. (10). Given the reaction $S_2O_8^{2-} + 3I^- \rightarrow 2SO_4^{2-} + I^- + I_2$ at room temperature, and the calculated rate values in Table 2B, determine the following:
   a. Order of the reaction with respect to $S_2O_8^{2-}$, (x), by using Runs 1 and 2
   b. Order of the reaction with respect to I^-, (y), by using Runs 1 and 3
   c. The overall reaction order
   d. The rate law for the reaction

2. Calculate the reaction rate constant *k* for Runs 1, 2, and 3 at room temperature. Calculate the average rate constants for the three runs.

3. Using the rate law, and the data in Table 3B, calculate
   a. The rate constants for determinations 3b and 3c
   b. 1/T [Convert units of temperature in Kelvin (T in °C + 273.15).]
   c. ln *k*

4. Plot 1/T vs. ln *k* to derive the value of activation energy, E_a.
   ($R = 8.3145\ Jmol^{-1}K^{-1}$)

STUDENT NAME	Brandon Breazeale
PARTNER	Josh Brown, Harrison Arcipcete
TA NAME	Lu Yu
DATE	2/26/13

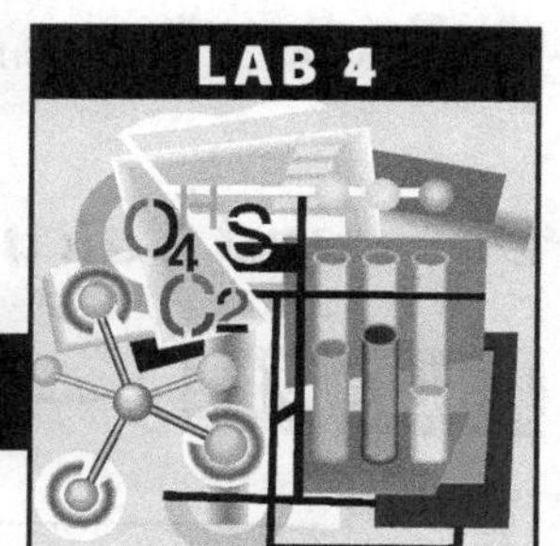

POST-LAB QUESTIONS

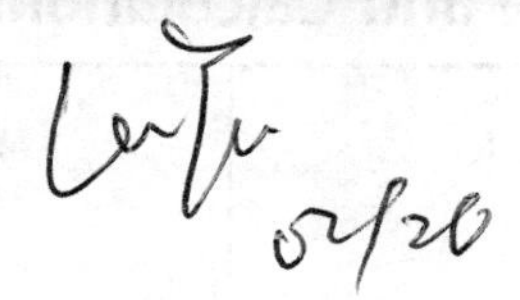

1. For Runs 1 to 3 of determining the rate of the reaction, we assume k is a constant. What assumption must be made to allow you to consider k to be a constant? [3 pt.]

2. If the initial concentrations of sodium persulfate and potassium iodide were set at 0.020 M and 0.060 M, respectively, what would be the approximate initial reaction rate? What time has elapsed when the first trace of blue color appears? [10 pt.]

3. What is the general equation that governs the relationship between reaction rate and temperature? [3 pt.]

4. Does the relationship between 1/T and $\ln k$ appear to be linear? Are three data points sufficient to justify the relationship in the equation for temperature effect? [4 pt.]

Data and Calculations

table 2B

Data and Calculations (Procedure 1)

MEASURE/ CALCULATION	RUN 1	RUN 2	RUN 3
$[KI]_i$			
$[Na_2S_2O_3]_i$			
$[(NH_4)_2S_2O_8]_i$			
Avg. reaction time, Δt (seconds)			
Reaction rate (M/sec) [15 pt.]			
Order of reaction: [6 pt.]		with respect to $S_2O_8^{2-}$	
		with respect to I^-	
		Overall ($x + y$)	
Rate law: [3 pt.]			
Rate constant, k [16 pt.]			
	Average:		

table 3B

Data and Calculations (Procedure 2)

RUN	3A	3B	3C
Temperature, T (°C)			
Reaction time, Δt (sec)			
Temperature, T (K)			
1/T, (1/K)			
Reaction rate (M/sec) [10 pt.]			
Rate constant, *k* [10 pt.]			
ln*k*			

NOTE: include the units for rate constant, *k*

Graph and activation energy: [20 pt.]

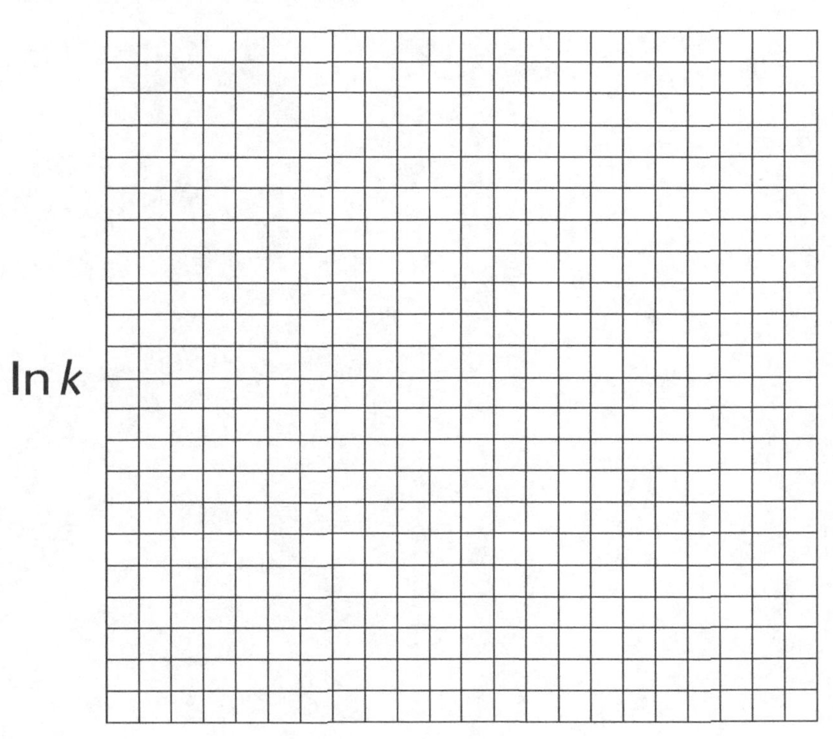

Last question: Describe one concept that you understand better as a result of this lab. Be specific. (10 pt.)

lab 5

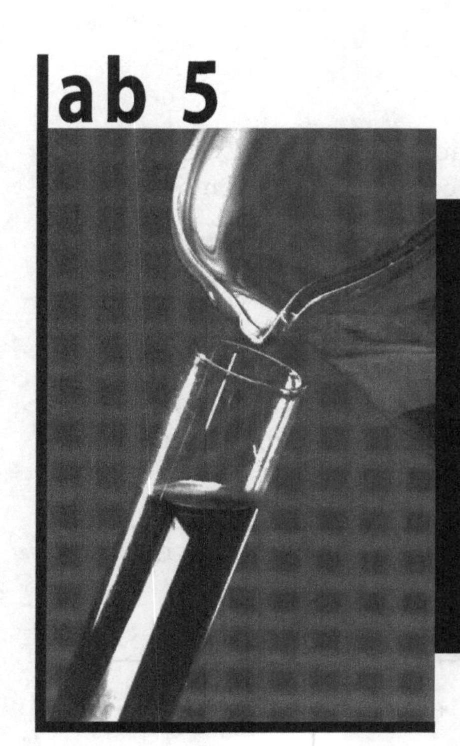

Equilibrium and Buffers

Chemically driven spontaneous reactions often "appear" to stop before the reactants are consumed completely. These reactions are reversible. Not only do the reactants react with themselves to give products, but the products react with themselves to give back the reactants. In other words, we say chemical equilibrium has been reached when the rates of forward and reverse reactions are equal but not zero.

Let us consider the following reaction:

$$CH_3COOH_{(aq)} + H_2O_{(1)} \rightleftharpoons H_3O^+_{(aq)} + CH_3COO^-_{(aq)} \quad \text{(Eq. 1)}$$

The rates of forward and backward reactions, respectively, are

$$\text{rate} = k[CH_3COOH][H_2O] \quad \text{(Eq. 2)}$$

$$\text{rate} = k_{-1}[H_3O^+][CH_3COO^-] \quad \text{(Eq. 3)}$$

At equilibrium the two rates are equal and the ratio of the rate constants $k/k_{-1} = K_c$, the equilibrium constant,

$$K_c = \frac{[H_3O^+][CH_3COO^-]}{[CH_3COOH][H_2O]} \quad \text{(Eq. 4)}$$

As the concentration of water remains essentially constant at 25°C,

$$K_c[H_2O] = \frac{[H_3O^+][CH_3COO^-]}{[CH_3COOH]} \quad \text{(Eq. 5)}$$

This new constant is the acid-ionization constant or the acid-dissociation constant (K_a).

$$K_a = \frac{[H_3O^+][CH_3COO^-]}{[CH_3COOH]} \quad \text{(Eq. 6)}$$

If we know the concentration of acetic acid (CH_3COO
corresponding K_a at 25°C, we can calculate the concentration o
CH_3COO^- ions.

Let us consider a solution that is 0.12 *M* in acetic acid at
1.7×10^{-5}). We will calculate the concentrations of acetic acid
CH_3COO^- at equilibrium. Before any ionization, the initial conc
the acid is 0.12 *M* in 1 L of water. The concentration of H_3O^+
ionization of water is negligible. If acetic acid ionizes to give x m
and x moles of CH_3COO^- ions then (0.12 – x) moles of undisso
acid remain at equilibrium.

CONCENTRATION (M)	CH_3COOH	H_3O^+	CH
Initial	0.12	~0	
Change	– x	+ x	
Equilibrium	0.12 – x	x	x

The equilibrium constant equation is

$$K_a = \frac{[H_3O^+][CH_3COO^-]}{[CH_3COOH]} = \frac{x^2}{0.12 - x} \approx \frac{x^2}{0.12} = 1.7 \times 10^{-5}$$

$$x = 0.0014\,M \qquad \text{(Eq. 7)}$$

Because the acid ionization is quite small, we can assume that x is negligible when compared to 0.12 *M*. Now, we substitute the value of x in the last line of the table to get the concentrations of acetic acid, H_3O^+, and CH_3COO^- at equilibrium as 0.12 *M*, 0.0014 *M* and 0.0014 *M*, respectively.

The $[H_3O^+]$ is a quantitative measure of whether the aqueous solution is acidic, neutral, or basic. This concentration tends to be rather small and, hence, it is more convenient to describe acidity in terms of pH where

$$pH = -\log[H_3O^+] \qquad \text{(Eq. 8)}$$

A neutral solution at 25°C, whose hydronium-ion concentration is 1.0×10^{-7} *M*, has a pH of 7.00.

For acidic solutions:	$[H_3O^+] > 1.0 \times 10^{-7}\,M$	$pH < 7$
For neutral solutions:	$[H_3O^+] = 1.0 \times 10^{-7}\,M$	$pH = 7$
For basic solutions:	$[H_3O^+] < 1.0 \times 10^{-7}\,M$	$pH > 7$

The figure on the inside back cover of this book illustrates the pH value of some common solutions.

Pure water shows weak auto-ionization to give:

$$H_2O + H_2O \rightleftharpoons H_3O^+ + OH^- \qquad \text{(Eq. 9)}$$

The equilibrium setup is

$$K_c[H_2O]^2 = [H_3O^+][OH^-] \qquad \text{(Eq. 10)}$$

As the concentration of water remains essentially constant at 25°C, the expression becomes

$$[H_3O^+][OH^-] = K_w = 1.0 \times 10^{-14} \qquad \text{(Eq. 11)}$$

where K_w is the ionic product of water

As water produces H_3O^+ and OH^- in equal numbers,

$$[H_3O^+] = 1.0 \times 10^{-7}\ M \quad \text{and} \quad pH = 7$$

$$[OH^-] = 1.0 \times 10^{-7}\ M \quad \text{and} \quad pOH = 7$$

and at 25°C,

$$pH + pOH = 14 \qquad \text{(Eq. 12)}$$

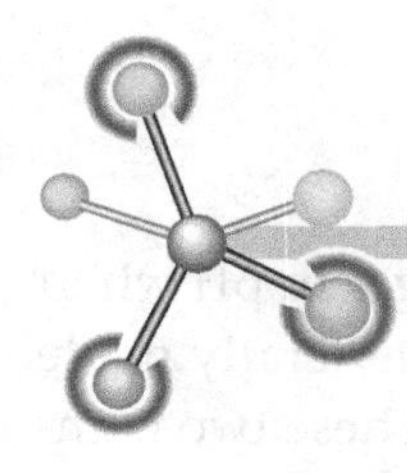

Ways to Monitor pH

Indicators

Acid-base indicators are usually weak acids and bases whose conjugate ions have different colors from the neutral molecules. The hydrolysis reaction shown below employs an indicator $H_{IN}D$:

$$\underset{\substack{\text{Weak acid} \\ \text{Color X}}}{H_{IN}D} + H_2O \rightleftharpoons \underset{\substack{\text{Conjugate base} \\ \text{Color Y}}}{{}_{IN}D^-} + H_3O^+ \qquad \text{(Eq. 13)}$$

At an optimal pH both the weak acid ($H_{IN}D$ color "X") and its conjugate base (${}_{IN}D^-$ color "Y") are present and the solution would have an intermediate color. Below this pH the solution would have weak acid predominantly and the color would be "X." Above this pH the solution would readily have the conjugate base and the persisting color would be Y. Two common indicators are phenolphthalein and methyl orange. Table 1 below lists their colors at certain pH values.

table 1

INDICATOR	pH VALUES	COLOR
Methyl orange	$pH < 3.1$	Red
	$3.1 < pH < 4.5$	Salmon pink
	$pH > 4.5$	Yellow
Phenolphthalein	$pH < 8.3$	Colorless
	$8.3 < pH < 10.0$	Very light pink
	$pH < 10.0$	Red

pH paper

pH paper is simply a multi-colored paper strip that is soaked in sample solution. The paper turns a different color (or set of colors) depending upon the pH of the solution. It provides a very coarse measurement of pH and is only sufficient for making simple determinations. It is too coarse a measurement for allowing comparisons between two samples.

pH Meter

pH can be measured using a pH meter that contains two electrodes. One electrode measures the activity of H^+ ions in the solution being tested, and the other measures the activity of H^+ ions in a reference solution that is pH 7. These are both measured as voltage. The voltage of the two electrodes is compared and from that the pH of the solution being measured is calculated. This calculation is done internally in the pH meter, thus you read the pH value listed on the display on the meter.

Buffers

A buffer is a solution possessing the ability to resist large changes in pH when limited quantities of an acid or a base are added. Buffers are generally made of two components: a weak acid and its conjugate base (salt). These two components are in excess relative to the small amounts of acid (H^+) and base (OH^-) added.

An example of a buffer would be the combination of acetic acid (CH_3COOH) and sodium acetate (source of CH_3COO^- ions, the conjugate base of CH_3COOH). The equilibrium system presented earlier in the lab is:

$$CH_3COOH_{(aq)} + H_2O_{(l)} \rightleftharpoons H_3O^+_{(aq)} + CH_3COO^-_{(aq)} \qquad \text{(Eq. 1)}$$

Le Chatelier's principle states that when a system in equilibrium is stressed by changing the concentration of one of the species present in the system, the equilibrium will shift to relieve the stress. In the acetic acid system above, this means that adding a small amount of excess acid will shift the equilibrium left when the extra H^+ ions react with CH_3COO^- to form CH_3COOH. Adding a small amount of excess base will shift the equilibrium to the right when the extra OH^- ions react with H_3O^+, resulting in a depletion of the hydronium ions. In either case, most of the excess H^+ and OH^- ions are taken out of circulation keeping the pH relatively stable. The concentrations of the remaining ions can be determined by the equilibrium constant of the reaction.

It is important to understand that a buffered solution does change pH with the addition of H^+ or OH^- ions, but such a change is much less than would occur if no buffer were present.

Balanced pH plays a very important role in the maintenance and regulation of biological processes. For instance, the blood has a constant pH of 7.4, made possible through the functioning of the kidneys and lungs that regulate certain chemical components in the blood. The carbonic acid-bicarbonate buffer is the most important buffer system in the blood. The reaction at equilibrium is as follows:

$$H_3O^+ + HCO_3^- \overset{K_1}{\rightleftharpoons} H_2O + H_2CO_3 \overset{K_2}{\rightleftharpoons} 2H_2O + CO_2 \qquad \text{(Eq. 14)}$$

We are interested in the change in the pH of the blood; therefore, we want an expression for the concentration of H^+ in terms of equilibrium constant and the concentrations of the other species in the reaction (HCO_3^-, H_2CO_3, and CO_2).

Applying the law of mass action for the left-hand reaction:

$$K_1 = \frac{[H_2CO_3]}{[HCO_3^-][H_3O^+]} \qquad \text{(Eq. 15)}$$

The reverse of the left-hand reaction represents the acid dissociation of carbonic acid:

$$K_a = \frac{1}{K_1} = \frac{[HCO_3^-][H_3O^+]}{[H_2CO_3]} \qquad \text{(Eq. 16)}$$

For the right-hand reaction:

$$K_2 = \frac{[CO_2]}{[H_2CO_3]} \qquad \text{(Eq. 17)}$$

Considering that both reactions occur simultaneously during equilibrium, we can solve both equations for $[H_2CO_3]$ and combine

$$[H_2CO_3] = \frac{[HCO_3^-][H_3O^+]}{K_a} = \frac{[CO_2]}{K_2} \qquad \text{(Eq. 18)}$$

Rearranging the equation to solve for equilibrium H_3O^+ concentration gives us:

$$[H_3O^+] = \frac{K_a}{K_2}\frac{[CO_2]}{[HCO_3^-]} \qquad \text{(Eq. 19)}$$

Taking negative log on both sides and using $K = K_a/K_2$, we get:

$$pH = pK - \log\frac{[CO_2]}{[HCO_3^-]} \qquad \text{(Eq. 20)}$$

The relationship expressed here is referred to as the Henderson-Hasselbach equation. It is used to calculate the pH of a given buffer or to calculate the molar ratio of salt to acid needed for a specific pH.

To calculate the pH of a acid/salt buffer we use the form:

$$pH = pK_a + \log\frac{[salt]}{[acid]} \qquad \text{(Eq. 21)}$$

LAB 5 EXERCISE

PURPOSE To understand the effect of buffer on the pH of a reaction.

PROCEDURE ❶ ***Caution:*** **The reagents used in this lab are corrosive to tissue. User must wear gloves at all times and the reagents need to be handled carefully.**

[NOTE: pH paper is extremely base and acid sensitive.]

Experiment 1

1. Use pH paper to determine the pH value of 0.1 *M* solution of HCl. Carefully measure out 1 mL of 0.1 *M* HCl in a test tube. Dip the pH paper into the test tube for 1 to 2 seconds. Remove the pH paper out and note the change in the color of the pH paper. Record the color change observed and the pH the color signifies. Discard the pH paper in the waste container.
2. Repeat this procedure with HOAc, NaOH, and NH_4OH.
3. Calculate a theoretical pH and compare your values.

table 1

Experiment 1			
SOLUTION (0.1 M)	pH (OBSERVED)	pH (CALCULATED)	pH PAPER COLOR (OBSERVED)
HCl			
HOAc			
NaOH			
NH_4OH			

NOTE: K_a HOAc = 1.8×10^{-5}; K_b (NH_4OH) = 1.8×10^{-5}.

Experiment 2

1. In a test tube or vial, add 2 mL of 0.1 *M* HOAc solution and 1 drop of methyl orange. Record your observations.
2. Then, to the test tube, add a small amount of solid NaOAc. Record your observations.

Experiment 3

1. In a test tube or vial, add 2 mL of 0.1 *M* NH_4OH solution and 1 drop of phenolphthalein. Record your observations.
2. Then, to this vial or test tube, add a small amount of solid NH_4Cl. Record your observations.

table 2

Experiments 2 and 3

SOLUTION	OBSERVATIONS
0.1 *M* HOAc + methyl orange	
0.1 *M* HOAc + methyl orange + NaOAc	
0.1 *M* NH_4OH + phenolphthalein	
0.1 *M* NH_4OH + phenolphthalein + NH_4Cl	

Experiment 4

[NOTE: Your TA will demonstrate the proper use of a pH meter.]

1. In a 50 mL beaker, mix 10 mL of 0.1 *M* HOAc and 10 mL of 0.1 *M* NaOAc. Swirl.
2. Using a pH meter or pH paper, determine the approximate pH value. Record.
3. Add 0.5 mL of 0.1 *M* of NaOH to the HOAc/NaOAc mixture. Swirl the solution.
4. Using a pH meter or pH paper determine the pH of the solution and record.
5. Repeat step 3, but add 0.5 mL 0.1 *M* HCl instead of NaOH.

Experiment 5

1. In a 50 mL beaker add 20 mL pure water.
2. Using a pH meter or pH paper, determine the approximate pH value. Record.
3. Add 0.5 mL of 0.1 *M* of NaOH to the water.
4. Using a pH meter or pH paper determine the pH of the solution and record.
5. Repeat step 3 but add 0.5 mL 0.1 *M* HCl instead of NaOH.

table 3

Experiments 4 and 5		
	0.1 M HOAc/0.1 M NaOAc	WATER
Initial pH		
pH after addition of 0.5 mL NaOH		
pH after addition of 0.5 mL HCl		

Experiment 6

Have one of the TAs check your calculations from pre-lab Question 3. Prepare your buffer and measure its pH.

Measured pH ______________________

Experiment 7

Predict the pH of the salt solutions listed in Table 4 using the equation listed in the introduction to the lab (*Hint: Does the salt contain a weak acid or a weak base?*). After you have calculated the pH of the solutions, obtain the solutions and measure the pH using the pH probe.

table 4

SALT SOLUTIONS (0.1 M)	pH CALCULATED	pH MEASURED
Potassium Acetate, KOAc		
Sodium Carbonate, Na_2CO_3		
Ammonium Nitrate, NH_4NO_3		
Magnesium Chloride, MgCl		
Potassium Bromide, KBr		

STUDENT NAME
PARTNER
TA NAME
DATE

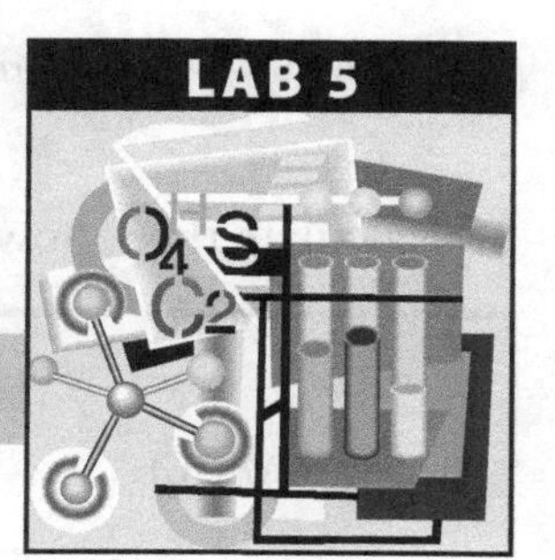

PRE-LAB ASSIGNMENT

1. Explain the following terms as they relate to this experiment by giving a suitable example. [25 pt.]

   a. Equilibrium

   b. Forward vs. reverse reaction

   c. Le Chatelier's principle

   d. Buffer

   e. pKa

2. Calculate the following quantities for each of the following solutions: [60 pt.]

	pH	pOH	$[H^+]$	$[OH^-]$
0.5 *M* HCl				
0.3 *M* KOH				
0.18 *M* NH_4Cl				

NOTE: K_b NH_3 = 1.8×10^{-5}

3. You have recently joined the Environmental Protection Agency (EPA) and your immediate task is to evaluate the acidity levels in the Onondaga Lake. For this purpose you need to prepare 100 ml of a 0.2 *M* acetate buffer, pH 5.0, starting from solid NaOAc and a 1 *M* solution of HOAc.

   [Formula weight of sodium acetate = 136, pKa of acetic acid = 4.77; HINT: Use the Henderson-Hasselbach equation (Eq. 21)] [15 pt.]

lab 6

Redox Potential

Oxidation-reduction reactions are reactions involving the transfer of electrons between species. Common processes involving oxidation-reduction include photosynthesis and combustion (the word "oxidized" is commonly used when referring to the process when atoms react with oxygen). Oxidation refers to the loss of electrons, while reduction refers to the gain of electrons. Often referred to as redox reactions, this combined term implies that an oxidation reaction will not occur without a reduction reaction happening at the same time. Each process by itself is called a "half-reaction," and by combining the two halves, we form the whole reaction. Half-reactions are written to include the electrons. The half-reaction for the oxidation (oxidation half-reaction) of zinc would be written:

$$Zn_{(s)} \rightarrow Zn^{2+} + 2\ e^-$$

Solid zinc forms a zinc ion through the loss of two electrons. Notice that in this half-reaction, we have a balance in both the mass and charge. The symbol "e^-" represents a free electron that may now be picked up by a second species for a reduction. Because nonmetals are easily reduced, one reaction possibility might be:

$$F_{2(g)} + 2\ e^- \rightarrow 2\ F^-$$

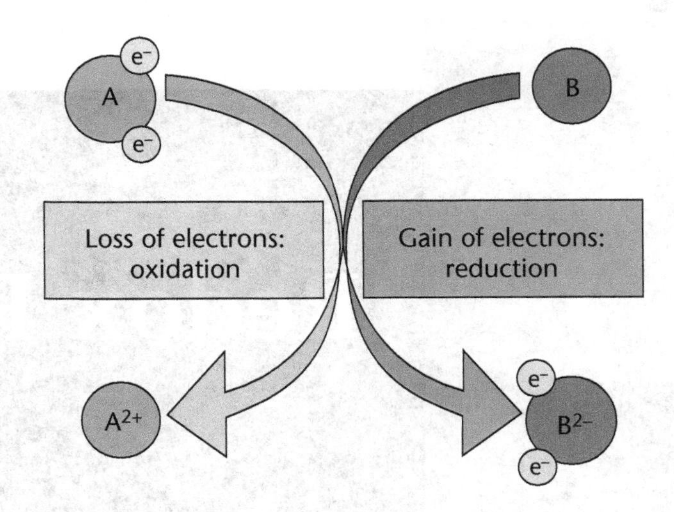

A fluorine molecule is reduced to two fluoride ions through the addition of two electrons. We can combine the two half-reactions to form a redox equation:

$$Zn_{(s)} \rightarrow Zn^{2+} + 2\ e^-$$
$$F_{2(g)} + 2\ e^- \rightarrow 2\ F^-$$
$$Zn_{(s)} + F_{2(g)} + 2\ e^- \rightarrow Zn^{2+} + 2\ F^- + 2\ e^-$$

or

$$Zn_{(s)} + F_{2(g)} \rightarrow ZnF_2$$

Individual components of these reactions are also referred to as *oxidizing or reducing agents*. The species that allows the oxidation to occur by receiving the electrons lost elsewhere (by a second species) is termed the *oxidizing agent.* In the above reaction $F_{2(g)}$ is the oxidizing agent because it picks up the electrons lost by the $Zn_{(s)}$. Oxidizing agents get reduced in the process by a *reducing agent.* $Zn_{(s)}$ is the reducing agent in this case, supplying the electrons to $F_{2(g)}$ for the reduction.

Because there can never be a reduction without an oxidation, the easiest way to calculate the energy change in a redox reaction is with a relative scale. The Standard Reduction Potential Table given in Appendix 1 is a compilation of this information.

The scale compares the ease of reduction for atoms (or ions) relative to a common atom. The common atom was chosen to be hydrogen (because it is very common and simple). The reduction of the hydrogen ion is located about midway down the chart and is assigned the reference value of 0.00 V.

Reduction potential is also an indication of a species' *affinity* for electrons. A substance with a positive reduction potential would be considered to have a greater affinity for electrons, while substances with a negative reduction potential have a lower affinity for electrons (these will more likely lose electrons).

The standard reduction table indicates the voltage, or potential, for many half-reactions. These voltages are measured at standard reaction conditions, with a temperature of 298 degrees Kelvin (or room temperature) and with a concentration of 1.0 moles per liter for each of the species. The potential of a reaction can be calculated using the reduction potentials.

Note, however, that this is a list of reductions. The reduction half-reactions that occur with the most ease are located at the top of the chart. Determining the oxidations that occur the most readily is determined from the bottom of the chart by reversing the given reaction. Using the information listed we can say that:

$$Li_{(s)} \rightarrow Li^+ + e^-$$

is a half-reaction with a high oxidation potential.

Using the chart, let us consider the combination of metal A and metal B. If A has a more positive reduction potential than B, then A will tend to gain electrons from B, becoming reduced and thereby oxidizing B. The greater the difference between the reduction potentials of A and B, the greater the activity with which electrons will move from B to A.

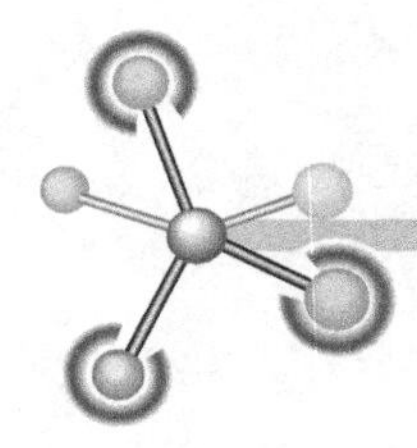

Metal Activity Series

An activity series is a list of chemical species based on their increasing or decreasing activity. Metals tend to react by losing one or more electrons. An activity series of metals is likely based on the ease with which this electron loss occurs. Metals listed high on a metal activity series are easily oxidized; metals in the low position on the activity series are not easily oxidized.

The activity of one metal (A) relative to another can be determined by placing metal A into a solution containing the cation of another metal (B). If metal A is more active and has a greater tendency to be oxidized, then metal A will displace B^+ cations from solution. One or more electrons is transferred and A goes into solution as A^+ while B^+ forms its elemental metal.

Once an activity series is created, it can be used to predict whether or not a redox reaction will take place. As an example, consider the following possible reaction:

$$Zn_{(s)} + Cu^{2+}_{(aq)} \rightarrow Zn^{2+}_{(aq)} + Cu_{(s)}$$

Does this reaction take place? If zinc metal is placed in contact with aqueous copper ion, will zinc displace copper ions from solution? Will Zn reduce Cu^{+2} or, equivalently, will Cu^{+2} oxidize Zn? As a general rule, a metal cation will be displaced from solution or its compounds by another element that is more reactive or higher on the activity series.

One of the reactants used in this lab is hydrochloric acid. Like all strong acids, hydrochloric contains hydrated protons, $H^+_{(aq)}$. When an active, easily oxidized metal is placed in contact with the acid, it will lose one or more electrons to the $H^+_{(aq)}$ producing metal ions and molecular hydrogen, $H_{2(g)}$. Bubbles of hydrogen gas will be visible on the surface of the metal. A less active metal will not produce this reaction.

A concept related to oxidation-reduction is the *ionization energy* of a substance. Ionization energy is the amount of energy required to remove an e^- from a neutral gaseous atom. Is there is a relationship between the activity of a metal and its ionization energy?

You will answer these and other questions by observing the relative activity of several metals and you will use your results to generate an activity series. Lastly, you will look up the ionization energy for each metal, and look for a relationship between the ionization energy and the activity of a metal.

Lab 7

Properties of Soap and Surface Tension

According to ancient Roman legend, soap gets its name from Mount Sapo. Here, fires were lit and animals were sacrificed in religious ceremonies. Rains mixed animal fat with the ashes from the fires, allowing a chemical reaction to occur. The continuing rain washed the product of this reaction into the nearby Tiber River, where women washing clothes in the river discovered that the foamy product made the clothes cleaner and with less effort than usual.

Soaps are made of molecules that are part oily and part water soluble. The oily part is hydrophobic and usually consists of aliphatic hydrocarbon chains; the water-soluble part often consists of ionic groups, either positive or negative charges with the corresponding counter ion. Because of this dual nature, these classes of molecules are also called *amphiphilic molecules*. With very few exceptions, these amphiphilic molecules prefer to be located at the surfaces of aqueous solution (or at the interface of air and the surface of the aqueous solution) because, loosely speaking, the oily part of the molecule is more "comfortable" when exposed to air than solvated in water. This surface-loving property of the molecules earns the molecules another name, *surfactants,* and also gives rise to their usefulness in a wide range of industries, including cosmetics, toiletries (yes, they share the same class of molecules, but not the same molecular structure), detergents, drug delivery, and others.

Besides getting to the surfaces of an aqueous solution, the oily part of the molecule also causes the molecules to form aggregates of a wide range of assembled structures (mostly micelles; others include vesicles, hexagonal, lamellar, and bicontinuous phases). Exactly which assembled structure will form depends upon the concentration and the molecular structure of the surfactants. Which part of the surfactant molecule do you suppose is exposed to water, and which part is buried in the aggregate away from water?

The chemical reaction that occurred at Mount Sapo is known as *saponification.* Fat or oil, also known as triglyceride, reacts with sodium hydroxide or potassium hydroxide (found in wood ashes) to form soap and a by-product known as glycerol. Triglycerides consist of three long-chain fatty acids linked by a three carbon chain (Fig.1).

The length of the fatty acid carbon chain can vary from 6 to 18 carbons long, although chain lengths of less than 10 carbons are considered unsuitable for soap production. The length of the carbon chain determines the hardness and solubility of the soap. Longer chains give harder, less soluble soaps, while shorter chains produce softer soaps. The base used in the saponification also affects the properties of the soap. Sodium hydroxide produces solid soaps; potassium hydroxide produces liquid soap.

The cleaning effectiveness of soap lies in two different properties. First, soap is a *wetting agent.* When a surface is contaminated by oil or grease, it becomes hydrophobic, or repellant to water. When water is placed on such a surface, it will not spread to evenly cover the surface. It will "bead up" as the water molecules are attracted to each other and repelled by the surface. The addition of soap to the water lowers its surface tension, allowing it to spread and more evenly wet the surface.

Second, soap is an *emulsifier.* It is well known that oil and water are immiscible. If you put oil and water together in a container and shake, drops of oil will be dispersed throughout the water. When you stop shaking the container, the oil drops will quickly coalesce and form a separate layer on top of the water. However, if you add an emulsifier (such as soap) to the mixture and shake, the oil will remain distributed throughout the water. Thus, soap allows water to dissolve grease and oil.

This ability of soap lies in the *amphiphilic* nature of its molecules; they contain both a polar, hydrophilic, "head group" (the salt of the carboxylic acid) and a non-polar, hydrophobic, "tail group" (the aliphatic carbon chain). Typically, an amphiphilic molecule is represented by a drawing in which the polar head group is shown as a circle, and the tail group is represented by a wavy line (Fig. 2).

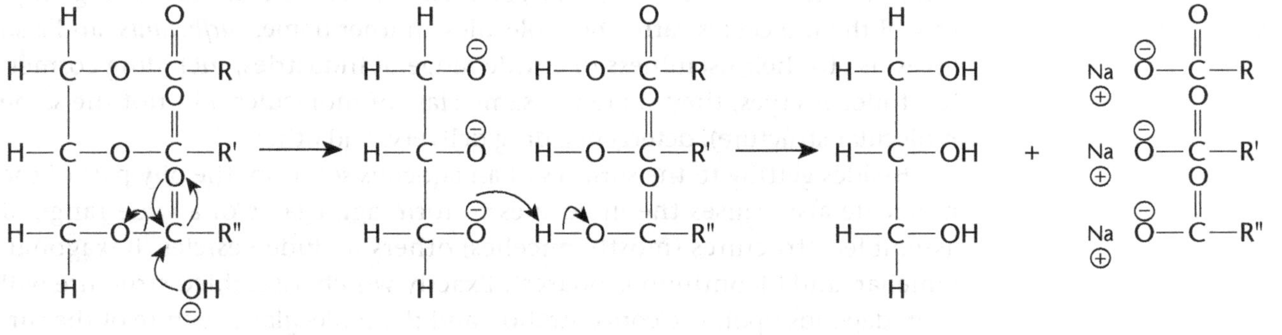

FIGURE 1. Saponification reaction of triglyceride with sodium hydroxide. The R groups of the triglyceride consist of aliphatic hydrocarbon chains, varying in length from 6 to 18 carbons (for soap manufacture, only fats with 12 to 18 carbon chains are used).

$^{-}O-\overset{O}{\overset{\|}{C}}(CH_2)_{15}CH_3$

Polar head group

Non-polar tail group

FIGURE 2. Structure of an amphiphilic molecule, showing the hydrophilic and hydrophobic portions of the molecule, and a schematic representation of the molecule.

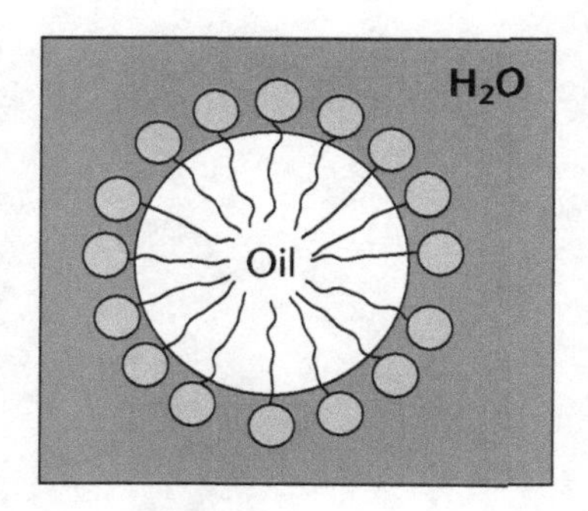

FIGURE 3. A micelle of oil and amphiphilic molecules in water is shown. The hydrophobic tails of the amphiphiles are dissolved in the oil, while the hydrophilic heads are dissolved in the surrounding water.

In a typical oil-in-water emulsion, drops of oil will be combined with the soap molecules, such that the hydrophobic tails of the soap are within the oil drop and the hydrophilic heads are outside the oil in the surrounding water (Fig. 3). This formation is known as a *micelle*.

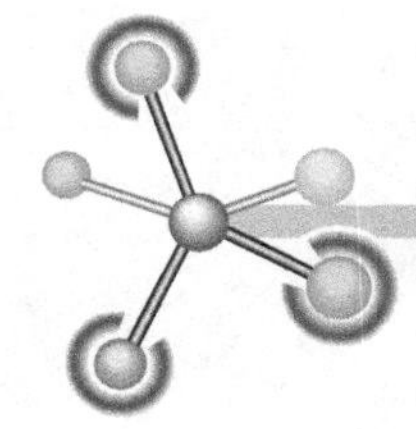

Surface Tension

Surface tension of water is caused by the water molecules at the surface being at a higher state of energy than the water molecules in the bulk of the water. A water molecule is at a lower state of energy when it is surrounded by other water molecules, forming the maximum amount of hydrogen bonds. When it is at the interface of air (or another medium) and water, it cannot form all the possible hydrogen bonds (imagine it). As such, the water molecules on the surfaces are being pulled into the bulk to achieve a lower state of energy and all the possible hydrogen bonds with other water molecules. Although this pulling force is balanced by the incompressibility of the liquid, when the shape of a volume of liquid water is allowed to change such as in a gravity-free space or when liquid water is thrown into someone's face (do not do that in the lab), the volume of water would tend to minimize its surface area so that the number of molecules on the surface is minimum. This minimization of the surface area gives rise to a tension on the surface of water, and thus the term *surface tension*. Note that for a cube and a sphere having the same volume, the sphere will have a smaller surface area.

Addition of surfactants to water will cause the assembly of a molecular layer of surfactant on the surface of the water. This molecular layer will reduce the surface tension of water to great extent. Changes in the surface tension in water can cause floating objects to move or to sink.

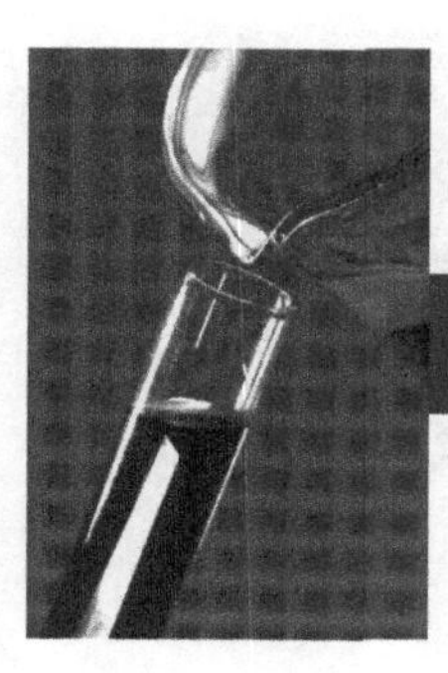

LAB 7 EXERCISE

PURPOSE To make soap and study properties of the soap, including dissolving oil into water and driving a paper boat across the surface of water.

PROCEDURE 1

Preparation of Soap

In this procedure you will make a bar of soap. This will be regular soap that can be used for washing, the same as you would use any soap bought in a store.

1. You will need a hot water bath. Put 2 to 3 inches of water in a large (600 mL) beaker, and heat on a hot plate to 70–80°C.
2. Place a 250-mL beaker on the balance and tare it (zero the balance).
3. Weigh 50 g of fat into the beaker.
4. Place the 250-mL beaker containing the fat into the water bath, but do not allow the water to enter the 250 mL beaker.
5. Heat, while stirring, until the fat is melted.
6. Add 5–35 mL of olive oil.

   [CAUTION: Do not heat the fat directly on the hot plate—it could ignite if overheated! Always use the water bath.]
7. Remove the melted fat from the water bath. Set this mixture aside to cool.
8. In a 150-mL beaker, weigh 6.25 g of sodium hydroxide.

   [CAUTION: Sodium hydroxide is very caustic—avoid skin contact. Also, when dissolving sodium hydroxide, the beaker will become very hot.]
9. Carefully add 17.5 mL of distilled water to the sodium hydroxide. *Stirring continuously to avoid the sodium hydroxide from clumping together.* Continue stirring until all of the sodium hydroxide is dissolved. Allow to cool. Both the oil and fat mixture and the sodium hydroxide solution should be at approximately 40°C for the next step. Cool further or reheat in the water bath as necessary to reach the desired temperature.
10. Carefully pour the sodium hydroxide solution into the fat and olive oil mixture while stirring continuously.
11. Continue stirring until the solution reaches a thick consistency (about the consistency of ketchup).
12. Immediately pour the solution into a mold (plastic weighing dish). Cover the mold with another weighing dish, and set it aside where it will not

be disturbed until next week. The soap will harden during this time, as the saponification reaction continues. After this the soap can be used.

13. Retain the soap left in the beaker for the next part of the procedure.

PROCEDURE 2

Examine the Effects of Adding Divalent or Multivalent Cations on Soap Solutions

You may have heard that hard water reduces the cleaning ability of soap. This is caused by the presence of divalent or multivalent cations (layman terms, minerals) in the water. When soap is used in hard water, multivalent cations replace the sodium cations in the soap molecules. The divalent or multivalent nature of these cations (typically calcium, magnesium and iron) has a much stronger binding to negative charges on the organic part of the soap molecules than the monovalent cations. Consequently, the cation and the anion do not separate and are not solvated in water. The net result is reduced solubility of soap in water.

1. Scrape about 1 g of the remaining soap from Procedure 1 out of the beaker and transfer it to a clean 250-mL beaker.
2. Add 100 mL of water.
3. Heat in the water bath, stirring continuously, until the soap is dissolved.
4. Label three test tubes: 1, 2, and 3, and add 5 mL of the soap solution to each.
5. In tube 1, add 3 drops of a 10% *calcium chloride* solution.
6. Place a stopper in the test tube (or hold your thumb over the opening), and shake vigorously. Record your observations in Table 1.
7. In tube 2, add 3 drops of 10% *ferric chloride* solution; shake and record observations
8. In tube 3, add 3 drops of 10% *potassium chloride* solution; shake and record the observations.

table 1

Observations for Effects of Mineral

TEST TUBE/ 250 mL BEAKER	REAGENT	OBSERVATIONS	
		SOAP	SYNTHETIC DETERGENT
1	Tap water	Cloudy water	cloudy, with massive bubble formation on surface
2	3 drops 10% $MgCl_2$	cloudier water with particles collecting on top of solution	milky appearance, fewer bubbles
3	3 drops 10% $FeCl_3$	Cloudy but tinted yellow	yellowy orange milky, frothy on top
4	3 drops 10% KCl	Cloudy, with thin layer of particles on top of solution	Clearer water, thick bubbling

9. Label three more test tubes, and add 5 mL of synthetic detergent solution to each.
10. Repeat the procedure as for the soap solution above.
11. Record your observations. Note differences between the soap solution and the detergent solution.

PROCEDURE 3

Observing Emulsification

1. Put 5 mL of distilled water in a test tube, and add 5 drops of olive oil.
2. Stopper the tube, or hold your thumb over the top, and shake vigorously.
3. Stop shaking the tube and immediately observe the appearance of the mixture.
4. Continue to observe the mixture over the next several seconds and record your observations.
5. Put 5 mL of your soap solution into another test tube, and add 5 drops of olive oil.
6. Stopper the tube and shake vigorously.
7. Stop shaking, observe and record your observations.

table 2

Observations for Emulsification	
Olive oil/water immediately after shaking	bubbles on top, cloudy water, greyish
Olive oil/water 15 sec after shaking	More milky, less bubbles, sediment on top
Olive oil/ water/soap immediately after shaking	lots of sediment, somewhat cloudy
Olive oil/ water/soap 15 sec after shaking	Solid particles on top, still milky coloring

PROCEDURE 4

Examining Surface Tension

1. Fill a dish with 1 to 2 inches of water.
2. On the surface, place several pieces of weighing paper in the shape of Figure 4.
3. Add 1 drop of water to the surface of the water and note your observations.

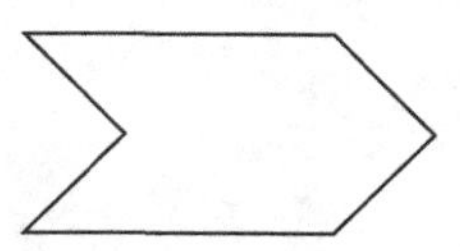

FIGURE 4

4. Repeat using a drop of salt water.
5. Repeat using the synthetic detergent.
6. Repeat using your soap.

table 3

Observations for Surface Tension

ADDED TO WATER	MOVEMENT OF PAPER PIECES
Water	Moved away from drop of water
Salt water	Moved inward
Detergent	Moved outward, away from drop
Soap	Nothing happened

STUDENT NAME	
PARTNER	
TA NAME	
DATE	

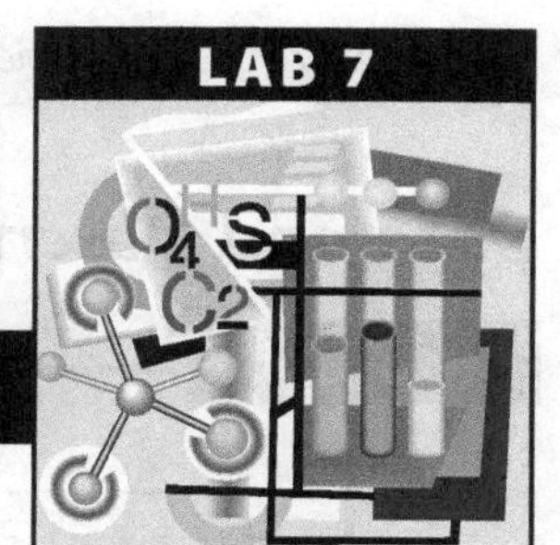

POST-LAB QUESTIONS

1. What are the charges on the metal ions of calcium and ferric chloride? Propose a reason why divalent or multivalent cations tend to reduce the water solubility (ask your TA for hints and suggestions), but monovalent cations do not have such a tendency. [6 pt.]

2. A beaker is half-filled with water. One drop of soap (let's say it is an anionic surfactant) is added to the edge of the surface of water in a beaker. Draw a schematic representation of the distribution and orientation of the soap molecules on the surface at the beginning (after 1 second) and after a longer time (5 hours). [20 pt.]

3. Suppose the soap you make does not work as well as a commercially available soap made by perfected reaction conditions. Describe what may have happened in making the soap that caused this substandard performance. [10 pt.]

Data and Observations

table 1

Observations for Effects of Mineral [32 pt.]

TEST TUBE/ 250 mL BEAKER	REAGENT	OBSERVATIONS	
		SOAP	SYNTHETIC DETERGENT
1	Tap water		
2	3 drops 10% $MgCl_2$		
3	3 drops 10% $FeCl_3$		
4	3 drops 10% KCl		

table 2

Observations for Emulsification [16 pt.]

Olive oil/water immediately after shaking	
Olive oil/water 15 sec after shaking	
Olive oil/ water/soap immediately after shaking	
Olive oil/ water/soap 15 sec after shaking	

table 3

Observations for Surface Tension [16 pt.]

ADDED TO WATER	MOVEMENT OF PAPER PIECES
Water	
Salt water	
Detergent	
Soap	

Last question: Describe one concept that you understand better as a result of this lab. Be specific. (10 pt.)

APPENDIX 1: STANDARD REDUCTION POTENTIALS IN AQUEOUS SOLUTIONS

HALF-REACTION	E_0(V)
$F_2(g) + 2e^- \rightarrow 2F^-$	2.87
$Co^{3+} + e^- \rightarrow Co^{2+}$	1.82
$Au^{3+} + 3e^- \rightarrow Au(s)$	1.50
$Cl_2(g) + 2e^- \rightarrow 2Cl^-$	1.36
$O_2(g) + 4H^+ + 4e^- \rightarrow 2H_2O(l)$	1.23
$Br_2(l) + 2e^- \rightarrow 2Br^-$	1.07
$2Hg^{2+} + 2e^- \rightarrow Hg_2^{2+}$	0.92
$Hg^{2+} + 2e^- \rightarrow Hg(l)$	0.85
$Ag^+ + e^- \rightarrow Ag(s)$	0.80
$Hg_2^{2+} + 2e^- \rightarrow 2Hg(l)$	0.79
$Fe^{3+} + e^- \rightarrow Fe^{2+}$	0.77
$I_2(s) + 2e^- \rightarrow 2I^-$	0.53
$Cu^+ + e^- \rightarrow Cu(s)$	0.52
$Cu^{2+} + 2e^- \rightarrow Cu(s)$	0.34
$Cu^{2+} + e^- \rightarrow Cu+$	0.15
$Sn^{4+} + 2e^- \rightarrow Sn^{2+}$	0.15
$S(s) + 2H^+ + 2e^- \rightarrow H_2S(g)$	0.14
$2H^+ + 2e^- \rightarrow H_2(g)$	0.00
$Pb^{2+} + 2e^- \rightarrow Pb(s)$	–0.13
$Sn^{2+} + 2e^- \rightarrow Sn(s)$	–0.14
$Ni^{2+} + 2e^- \rightarrow Ni(s)$	–0.25
$Co^{2+} + 2e^- \rightarrow Co(s)$	–0.28
$Tl^+ + e^- \rightarrow Tl(s)$	–0.34
$Cd^{2+} + 2e^- \rightarrow Cd(s)$	–0.40
$Cr^{3+} + e^- \rightarrow Cr^{2+}$	–0.41
$Fe^{2+} + 2e^- \rightarrow Fe(s)$	–0.44
$Cr^{3+} + 3e^- \rightarrow Cr(s)$	–0.74
$Zn^{2+} + 2e^- \rightarrow Zn(s)$	–0.76
$Mn^{2+} + 2e^- \rightarrow Mn(s)$	–1.18
$Al^{3+} + 3e^- \rightarrow Al(s)$	–1.66
$Be^{2+} + 2e^- \rightarrow Be(s)$	–1.70
$Mg^{2+} + 2e^- \rightarrow Mg(s)$	–2.37
$Na^+ + e^- \rightarrow Na(s)$	–2.71
$Ca^{2+} + 2e^- \rightarrow Ca(s)$	–2.87
$Sr^{2+} + 2e^- \rightarrow Sr(s)$	–2.89
$Ba^{2+} + 2e^- \rightarrow Ba(s)$	–2.90
$Rb^+ + e^- \rightarrow Rb(s)$	–2.92
$K^+ + e^- \rightarrow K(s)$	–2.92
$Cs^+ + e^- \rightarrow Cs(s)$	–2.92
$Li^+ + e^- \rightarrow Li(s)$	–3.05

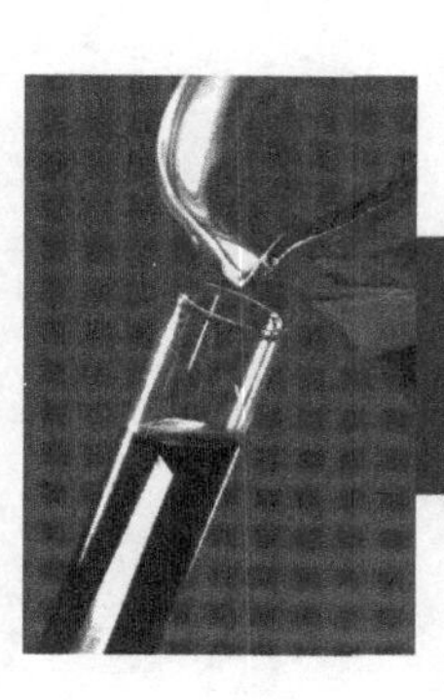

APPENDIX 2: PERIODIC TABLE OF THE ELEMENTS

1	2	3	4	5	6	7	8	9	10	11	12	13	14	15	16	17	18
1 H 1.00794																	2 He 4.002602
3 Li 6.941	4 Be 9.012182											5 B 10.811	6 C 12.0107	7 N 14.00674	8 O 15.9994	9 F 18.9984032	10 Ne 20.1797
11 Na 22.989770	12 Mg 24.305											13 Al 26.981538	14 Si 28.0855	15 P 30.973761	16 S 32.066	17 Cl 35.4527	18 Ar 39.948
19 K 39.0983	20 Ca 40.078	21 Sc 44.955910	22 Ti 47.867	23 V 50.9415	24 Cr 51.9961	25 Mn 54.938049	26 Fe 55.845	27 Co 58.933200	28 Ni 58.6934	29 Cu 63.546	30 Zn 65.39	31 Ga 69.723	32 Ge 72.61	33 As 74.92160	34 Se 78.96	35 Br 79.904	36 Kr 83.80
37 Rb 85.4678	38 Sr 87.62	39 Y 88.90585	40 Zr 91.224	41 Nb 92.90638	42 Mo 95.94	43 Tc (98)	44 Ru 101.07	45 Rh 102.90550	46 Pd 106.42	47 Ag 107.8682	48 Cd 112.411	49 In 114.818	50 Sn 118.710	51 Sb 121.760	52 Te 127.60	53 I 126.90447	54 Xe 131.29
55 Cs 132.90545	56 Ba 137.327	57 La 138.90547	72 Hf 178.49	73 Ta 180.94788	74 W 183.84	75 Re 186.207	76 Os 190.23	77 Ir 192.217	78 Pt 195.078	79 Au 196.96655	80 Hg 200.59	81 Tl 204.3833	82 Pb 207.2	83 Bi 208.98040	84 Po (209)	85 At (210)	86 Rn (222)
87 Fr (223)	88 Ra (226)	89 Ac (227.03)	104 Rf (261)	105 Ha (262)	106 Sg (263)	107 Ns (262)	108 Hs (265)	109 Mt (266)	110 (269)	111 (272)	112 (277)		114 (289) (287)		116 (289)		118 (293)

58 Ce 140.116	59 Pr 140.90765	60 Nd 144.24	61 Pm (145)	62 Sm 150.36	63 Eu 151.964	64 Gd 157.25	65 Tb 158.92534	66 Dy 162.50	67 Ho 164.93032	68 Er 167.26	69 Tm 168.93421	70 Yb 173.04	71 Lu 174.967
90 Th 232.0381	91 Pa 231.03588	92 U 238.0289	93 Np 237.05	94 Pu 244	95 Am (243)	96 Cm (247)	97 Bk (247)	98 Cf (251)	99 Es (252)	100 Fm (257)	101 Md (258)	102 No (259)	103 Lr (262)